Science
NEXT

Science
NEXT

Innovation for the Common Good from the Center for American Progress

Edited by Jonathan D. Moreno and Rick Weiss

Bellevue Literary Press
New York

First published in the United States in 2009 by
Bellevue Literary Press, New York

FOR INFORMATION ADDRESS:
Bellevue Literary Press
NYU School of Medicine
550 First Avenue
OBV 640
New York, NY 10016

NOTE:
A somewhat different version of "Biopolitics and the Quest for Perfection"
appeared as "The Clone Wars" in *Democracy*, Issue 5, 2007; "Your DNA's
Spitting Image" appeared as "What You Should Know Before You Spit Into
That Test Tube" in the *Washington Post*, July 20, 2008.

This book was published with the generous support of
Bellevue Literary Press's founding donor the Arnold Simon Family Trust,
the Bernard & Irene Schwartz Foundation
and the Lucius N. Littauer Foundation.

Library of Congress Cataloging-in-Publication Data

Book design and type formatting by Bernard Schleifer
Manufactured in the United States of America
FIRST EDITION
1 3 5 7 9 8 6 4 2
ISBN 978-1-934137-18-5-TP

Contents

Science
NEXT

Foreword

L IKE MANY AMERICANS, UNTIL I WAS FACED WITH MY OWN HEALTH
problems, the importance of science and technology for our way
of life seemed somewhat abstract to me, helpful in some global way but
not particularly relevant to me. And then, faced with a cancer that
might have taken my life swiftly and painfully only a generation
before, I saw firsthand how relevant it really was. Because of the investments of previous generations, modern medical science was able to
help me look forward to more enjoyable and productive years. One of
the reasons I have dedicated some of my bonus time to improving our
health-care system is so that everyone in need can realize the fruits of
innovation, as I have.

Yet America's creative genius is faced with more challenges than
ever. It cannot prosper without the public's confidence and support.
Even Thomas Edison, who we often think of as the classic lone inventor, realized that as technology modernized, it required teams to develop and disseminate. Nearly 75 years after Edison's light bulb, Jonas
Salk worked with a group at a university research lab to develop his
polio vaccine. And we know what a magnificent effort was required to
put an American on the moon nearly 100 years after Edison's first
incandescent bulb.

Original ideas are precious commodities. Acted upon, they shape
everything from the well-being of our loved ones to our national security to the ways we eat, work, and play. And they can stretch farther
than that: they are a window into the entire universe.

Our country has been fortunate to have so many leaders who
understood the need to nurture science. That vision is needed now
more than ever. Science is more expensive than ever, the problems we

face are more complex than ever, and we are faced with global competition on an unprecedented scale. Our friends, competitors, and adversaries also appreciate that science-based knowledge holds one of the keys to the human future. And, of course, the information revolution has helped scientists collaborate across obstacles of space and time that once held them back. We should celebrate the rising tide of knowledge that will help lift all boats, but we should not fool ourselves into thinking we have some unassailable right to be the lead vessel.

I'm excited about *Science Next* because in its pages I sense visions of the future that combine knowledge with a concern for justice, marrying what we can be intellectually with what we can be morally. Innovation is not simply an abstract victory of knowledge; it is not just the research that gave me years to live; the next science can advance human flourishing and serve the common good. That's the kind of world I want to leave for my children, and for yours.

—ELIZABETH EDWARDS

Introduction: Time for Science to Reclaim its Progressive Roots

ELCOME TO *SCIENCE NEXT*, A COLLECTION OF SOME OF THE MOST exciting and far-reaching ideas about innovation for a new American century.

The writings in this volume emerged from a literary experiment that has been evolving during the past year on the virtual and paper pages of *Science Progress* (www.scienceprogress.org), which is a project of the Center for American Progress, a Washington, D.C.–based policy-research institute. The mission of *Science Progress* is to provide an opportunity for scientists and non-scientists to share ideas about ways that scientific and technological innovation can contribute to human flourishing.

Given its genesis in a Washington think tank, the *Science Progress* conversation focused first on "inside the beltway" policymakers—a much-maligned but invaluable American species. Derided in the vernacular of Capitol Hill as "wonks," these public servants and their minions are burdened with the enormous responsibility of translating the nation's collective knowledge and wisdom into practical, political, and economic action.

We at *Science Progress* have grown increasingly inspired, though, by the range of smart ideas outside those conventional circles and by the public hunger to become more a part of the process of bringing the art of science to good governance. With *Science Next* we take the conversation to a new level, and invite you to be part of it. After all, "wonk" spelled backwards is "know." And it is knowledge—including public knowledge and understanding of science as an engine of progress—that will reveal solutions to today's most pressing problems, including climate change, energy independence, and national security.

The phrase "science progress" is, arguably, a bit awkward. Some would say it is redundant; others, less sanguine about where science is going, might call it contentious. But we who have been cultivating the pages of *Science Progress* find the construction provocative in the best sense of the word. It reminds us that we are the inheritors of the Enlightenment's confidence in the possibility of improving the human condition—a possibility predicated on values of individual freedom, social equality, and democratic solidarity, and one that values reason as superior to dogma or blindly "received wisdom." From this standpoint, scientific inquiry is the paradigmatic exercise of Enlightenment values.

You got a problem with that? Well let's go at it, because one of the things we love about science is that it is nothing if not argumentative. Both as a way of thinking and as a wellspring of novel ideas and products, science is a tumultuous truth-seeking process and even further, we contend, a revolutionary force for human liberation. This understanding of science as progressive does not deny that the power of science may be misused. Nor does it exclude the importance of other sources of inspiration or belittle the need for guidance and even regulation to ensure that the products of our progress are distributed fairly. But it does assert that the core values of science are democratic and antiauthoritarian. And it reflects a philosophical commitment to perpetual change and improvement over certainty and stasis.

The very words "science" and "progress" took on their modern meanings in the nineteenth century, and it should not be surprising that they came of age around the same time. It was an era in which microscopes and telescopes were drilling down and up into nature, while stethoscopes were revealing the body's mysterious inner space. Systematic investigation involving the careful manipulation of isolated variables was beginning to prove itself superior to mere observation, speeding the shift from mere anecdote to real evidence. The possibilities that could emerge from human insight were beginning to seem endless.

Science as progressive, however, boasts philosophical and political skeins stretching much further back into the American historical experience. Francis Bacon's utopian *New Atlantis* is often credited as being the first literary work to express the modern idea of progress in terms of advancing science and technology. It was a vision that was to have a profound effect on later seventeenth-century thinkers, including those who provided the intellectual justification for the American Revolution. For all the founders' disagreements, they shared the conviction that the new nation's promise was necessarily bound up with its innovative genius. Even those bitter rivals Jefferson and Hamilton were of one mind as they made their synergistic contributions to America's identity as a nation dedicated to modernity: Jefferson through the patent statute and Hamilton by laying the foundations for history's most successful capitalist economy, which together have so rewarded and nourished inventiveness.

It is no coincidence that so many of the concepts at the very heart of how America has come to understand itself—ideas such as the frontier and the West—demand an experimental attitude in grappling with novel challenges. The optimistic "can do" spirit; the approval of bigness, boldness, and adventure; the lure of "the road"—all are associated with this sensibility and are at the heart of our veneration of this country's great inventors, people like Benjamin Franklin, Thomas Edison, Jonas Salk, and Bill Gates. We hold these truths of perseverance and perspicacity to be, if not self-evident, at least within our grasp.

Even as America's western frontier has vanished, the pioneer spirit and the virtues and values associated with it have maintained their powerful hold over the American psyche. Inspired by that vision, Americans have repeatedly heeded the call to cross new and ever more challenging frontiers—including those well beyond the comforts of our cozy planet. Indeed, few government initiatives have been so wildly successful in capturing the public imagination as the space program of the 1960s, which explicitly drew upon the American frontier spirit. "[W]e stand today on the edge of a New Frontier," John F. Kennedy exhorted in 1960 as he clinched the Democratic nomination for president. "Beyond that frontier are the uncharted areas of science and space, unsolved problems of peace and war, unconquered pockets of ignorance and prejudice, unanswered questions of poverty and surplus."

Generations of Americans have come to characterize the United States itself as an experiment, a romantic and visionary theme compatible in orientation with pragmatist philosophers and early progressives. In this view, the only sure path to social and scientific advancement is as an iterative process of hypothesis, systematic experimentation, and data-gathering, followed by reform in light of experience. That the human condition can and should be improved by any means necessary—whether through government or private enterprise or some combination of the two, but with government as the ultimate guarantor of the public interest—has come to be the essence of progressivism, ever grounding those alleged improvements in the best possible evidence.

America's emergence as a nation of perpetual progress is all the more impressive given that this historical theme is not an inherent element of Western culture. The Greeks tended to think of their own time either as inferior to the mythical Golden Age or as part of a cycle of advance and decline. Imperial Romans saw themselves as in stasis since the establishment of the empire. Medieval Roman Catholic thinkers largely gave up on worldly progress in favor of spiritual improvement while awaiting Armageddon.

And perhaps reflecting these cautious and frankly depressing roots, the conjunction of science and progress in the modern era has not always been welcomed as an unalloyed good. Just as the words' modern meanings were coming into consciousness there were also the first signs of alarm, in a tradition that began famously with Mary Shelley's *Frankenstein* and continues to exert a powerful hold on popular culture today. Taken to an extreme, this view holds that far from being a guarantor of progress (a promise that even progressives could not reasonably make), the potentially inhumane and even dehumanizing drift of science threatens the furtherance of progress itself.

One common criticism of progressive science policy is that it naively adopts an instrumental view of science without reflection on the goals of innovation. At *Science Progress*, we appreciate that progressives have too often appeared to worship at the altar of change, and we reject the notion that a philosophy of innovation must be dumb to moral values. As you will see, *Science Next* considers ends as well as means, moral values as well as instrumentalities, as it explores the places where new ways of thinking can inform good governance.

Similarly, at the risk of invoking a hackneyed reference to spirituality, we also believe that science occupies an exalted dimension, that the growth of reliable knowledge is in effect an expansion of consciousness. Science may not be the only path to a greater grasp of reality, but it makes a unique contribution to enhanced understanding of the cosmos and our place within it. To be sure, science is a social enterprise, conducted in the service of the metaorganism—We the People—that is funding the work, and it bears a profound responsibility to respect its roots. But to distort the process of inquiry through the imposition of political or religious filters amounts to a narrowing of vision, a corruption of imagination, and a threat to our freedom as beings endowed with intellect.

One need not hark back to Copernicus or Galileo to see how such distortions can affect the arc of progressive science. It seems to many Americans that in recent years the respect for evidence and the spirit of open inquiry has been undermined and even sabotaged for the sake of short-term political advantage. The complex machinations of the American electoral system have recently placed the United States under new management, and there is reason to hope that science may once again find a more respected place at the policymaking table. It should be obvious to all that it is in the nation's long-term interest to have the best evidence available—evidence that in many cases only science can provide—to foster commercial innovation, economic growth, energy efficiency and environmental stewardship, educational advancement, military defense, and the best possible array of intelligence options.

In the twenty-first century, more than ever, it is no exaggeration to assert that only free and rigorous inquiry, and not authoritarian dicta, can provide the reliable information required for our physical survival. Open inquiry is also the best ticket to developing the tools that will allow us to fulfill our moral obligations to others in need, and to the planet itself. Perhaps most important, progress in science is essential for a continued sense of our national purpose as participants in a historic experiment in freedom and self-governance, as one people joined by a common future rather than a common past, a future we cherish not only for ourselves but for the sake of the generations of Americans to come.

Now we invite you to dip into *Science Next*, where our future may be written.

—JONATHAN D. MORENO AND RICK WEISS

America Can Do It: Reinventing the Soviet Satellite Stimulus Plan

by Vinton Cerf

REFLECTING FORWARD ON OUR NATION'S INCREDIBLE ABILITY TO RESPOND swiftly to complex scientific challenges, one can't help but begin with the shock of the Soviet Union's Sputnik launch a little more than a half-century ago. Suddenly, there was a new star in the firmament, its radio signal sounding like the ominous ticking of a clock toward Armageddon. If the Soviets could put a satellite in orbit, then they could potentially launch nuclear-tipped intercontinental ballistic missiles in the future.

Just over a decade later, the United States combined the basic science, the requisite technologies, and the financial and policymaking wherewithal to carry Neil Armstrong and Buzz Aldrin to the Earth's moon. This stunning sequence of events and actions spells out in forceful and compelling terms the ability of the United States to marshal its resources to respond to national and international challenges. That we were able to do so is an inspiring lesson that should be applied to the challenges we now face in the twenty-first century.

Fortunately for humankind, our nation's advanced science and technology can be harnessed to respond to a very different set of threats than those posed by the Cold War. Even better, all of humanity itself is our common ally in this quest.

Vinton Cerf is vice president and Chief Internet Evangelist for Google, and was a key architect of the Internet.

Today, quicksilver globalization powered at unprecedented speed by the information-technology revolution leaves our planet choking on the fumes of rapid global economic growth and all the attendant ills of global climate change. Yet the same forces that are transforming the global economy and the very globe itself also enable each of us to be part of the solution.

Modern communications technologies invite us to think of cooperation rather than competition as the venue for new ideas and new wealth. It is widely recognized that while material assets such as factories provided the leverage for the creation of value in the industrial age, information is the new coin of the realm. And, unlike industrial infrastructure, which could benefit only one user at a time, access to information is not a zero-sum game.

As John F. Kennedy—the president who was, arguably, elected by Sputnik—liked to say, "A rising tide lifts all boats." As we look to solve global problems we need to take a more imaginative, less adversarial approach to generating the products and arrangements that will make for a more livable planet. Our national response to Sputnik fifty years ago should inspire us today to transcend national boundaries and move beyond simple competition as the framework of human achievement.

Global climate change is a planetary threat that the United States cannot do less than meet head-on with the same kind of determination and leadership that placed us on the moon. Global warming is our twenty-first-century Sputnik. Former Vice President Al Gore has been tireless and immensely persuasive in his efforts to draw attention to this problem. Yet it is time for scientists, technology leaders, financiers, and public policy makers to take the same kind of concrete, swift steps embraced by our country fifty years ago—steps that can result in a new flourishing of creativity and ingenuity emblematic of great scientific endeavors.

Remember, sweeping scientific inquiry informed by smart policy-making carried us to the moon, but the many ancillary results of that mission are part and parcel of our world today. New materials needed for re-entry into the atmosphere and for protecting human life during space walks found their way into commercial products. Significant computing power was brought to bear in the design and planning of space systems and missions and the analysis of rocket engine performance.

Or consider the command and control of the complex Apollo missions. Computers developed for those missions led directly to the develoepement of the Advanced Research Projects Agency Network, or ARPANET, and the succeeding Internet. Management practices for complex systems found their way from the space program into the private sector, enhancing productivity and scalability of enterprises.

A national focus on innovation can nurture and cultivate the best traits of a society and its individuals. Indeed, at age fifteen and already a science-fiction junkie, I would benefit directly from the enrichment programs stimulated in large measure by the Sputnik launch. Introduced to computers in 1958 via the Semi-Automated Ground Environment tube-based computer at System Development Corp. in Santa Monica, California, I found myself using computers at the University of California, Los Angeles, while a senior in high school and taking every computing course I could at Stanford University as an undergraduate student.

In 1965 I went to work for International Business Machines Inc. as a systems engineer and returned to graduate school at UCLA where I ended up working on the ARPANET project funded by the agency that was formed in response to Sputnik. I graduated and joined the faculty of Stanford University and, together with Robert E. Kahn, designed the basic architecture and protocols of the Internet.

This confluence of events in my life, and in the lives of many other American scientists and engineers, is no accident. Fifty years ago the United States rose to the challenge. Similarly, resources dedicated today to the challenge of global warming will ensure innovation continues to flourish across our planet, no doubt with unimagined ancillary technological and economic benefits.

In fact, many of the steps that can be taken to respond to the serious dislocations that global warming will cause make eminently good independent economic sense. The development of high-mileage, internal combustion engine cars, or alternative clean-energy vehicles such as hybrid automobiles or all-electric cars, would reduce pollution and dependence on oil imports. The research needed to achieve this objective could be led by the American automobile industry and even subsidized by a civilian equivalent to the Defense Advanced Research Project Agency (DARPA). And just imagine what other benefits would flow from a dedicated wave of research into lighter and stronger materials, more efficient and lighter-weight batteries or fuel-cell systems, and alternative fuel sources.

The same multiplier effect holds for other scientific endeavors into a variety of green technologies, among them reduced-energy light sources; more efficient heating and cooling designs and technologies; better mass-transit systems; higher speed and more widely available communication services to support working from home; more effective traffic-control systems to reduce congestion and wasted energy; improved desalinization methods to cope with the loss of fresh water from mountain snowpack and glaciers or underground aquifers; and global power grids to transport electricity from areas of excess to areas of need.

Indeed, a post-Sputnik-like response to climate change would inevitably spur innovation in seemingly unrelated terrain brought on by global warming, such as preparing for the likely increase in diseases caused by the side effects of severe weather and storm surges. Moreover, successful results could become the basis for valuable international economic trade because the uses for these ideas are not bound to their domestic origins.

The National Science Foundation, for example, could accelerate the development of curricular material to emphasize science, mathematics, and engineering in the interest of responding to climate change. The Internet can be used to disseminate this material and to share information globally to speed the research that is needed. Prize programs could be established to encourage research and experimentation in areas of specific need.

Considering the consequences of not responding to this planetary challenge, it seems inescapable that the United States can and must take a leadership role. Other countries may soon exceed, in absolute terms, our contribution to global warming, but we consume more resources and generate more greenhouse gases per capita than any other country. Besides, it is in our, and everyone's, best interest to develop alternative technologies and share them widely.

We responded effectively to Sputnik, and this is even more important. We can do it again.

TENDING TO TERRA

It goes without saying, yet is rarely acknowledged or deeply considered, that for now, at least, we have but one home, a little blue-green marble in the vast expanse of space. Yet a visitor from some other celestial orb might be shocked to see how we've soiled our precious nest.

Happily, collective consciousness of the big impact that our countless small actions have been having on the Earth, its oceans and our atmosphere is beginning to rise, and not a decade too soon.

Some are quick to note that science and technology were central culprits in creating the predicament we're in today, and there is no denying that without the leverage gained with modern science, humanity would have been hard-pressed to do all the damage it has done to date. But the same skill sets that wrought ubiquitous seeding of PCBs and DDT, global climate change, and shrinking biodiversity also have the potential to help us solve our problems and perhaps rehabilitate our reputations as good and upright members of the ecosystem.

Here we offer some positive tales of humankind's potential to redeem its relationship with its mother.

Solar Rays to the Rescue

THE SPRING OF 2008 MARKED THE FIFTIETH ANNIVERSARY OF A LITTLE-known but seminal event in U.S. history: the launch of this nation's first solar-energy system. And we really do mean "launch." In March 1958 the fledgling technology for making electricity from sunlight was embedded in a grapefruit-sized satellite that American engineers flung into orbit as part of America's rapid response to the Soviet Union's launch of Sputnik. Sunlight was to power the U.S. satellite's transmitter, which would beep-beep-beep its presence to Earthbound receivers, proving that the United States had a solid stake in the new frontier of space.

As recounted by Lawrence Kazmerski of the Department of Energy's National Renewable Energy Laboratory, there was some debate among the engineers designing that satellite. Some just wanted to rely on a battery. As it worked out, both technologies were incorporated. The battery worked fine . . . for less than three weeks. After it died, the solar cells kicked in and the satellite kept beeping and beeping—for so long, in fact, that the satellite became something of a pest, tying up that Naval Research Laboratory radio frequency for nearly seven years before finally petering out in 1964.

Lesson learned: next time, include an on-off switch.

With those robust beginnings, one might expect, or at least hope, that by now solar energy would be everywhere. But of course it is not (though it is in every one of the many satellites now circling the globe). Solar power today accounts for just a fraction of one percent of U.S. electricity generation, despite our early venture into the technology and despite the Manhattan Project–like call by President Jimmy Carter to get at least 20 percent of our electricity from solar way back in the 1980s.

The reasons for this failure are many. Financial incentives promulgated by Carter were dismantled by his successor, President Ronald Reagan, who also gutted Carter's clean-energy program. The technology has, until recently, remained frustratingly expensive. And shortages of silicon, a key ingredient for most photovoltaic systems, put a crimp on emerging businesses in the early part of the new millennium.

Now a perfect storm of the best kind appears to be gathering. New technologies are increasing efficiencies and lowering costs. Fossil-fuel prices are high—oil imports account for almost half of the nation's total trade deficit—making it easier for alternative energies such as solar to prove their cost-effectiveness. Climate change adds new incentives for the nation (and indeed, the world) to shift away from carbon-emitting power sources. And here in the United States, at least, huge expanses of territory are constantly awash with sunlight. Indeed, the American Southwest is a veritable Saudi Arabia of sun.

And yet it is not out of the question that the United States may squander this opportunity, according to scientists, investors, and politicians. And as with so many issues in these treacherous times, the factor that stands to make all the difference one way or the other is "political will."

That's a general term, of course, and one that is thrown around a lot in Washington. But happily, there seems to be a growing consensus as to what "political will" really means in the realm of renewable and sustainable energy, and in particular solar energy. Here are the main elements of consensus increasingly expressed by a wide range of stakeholders:

Institute a cap and trade system that places a price on carbon emissions commensurate with their damage. Everything else follows from this. Until carbon is accurately valued—or devalued—the energy playing field will remain too crooked to stand on.

Create a National Renewable Production Standard. Many states set enforceable goals for themselves for what percentage of their energy production should come from renewable sources. But unless the nation does the same across the board, it will be impossible to have coordinated transmission among regions that are rich in different resources (most notably, sun in the Southwest and wind in the Midwest). In a November 2007 report, for example, the Center for American Progress called for 25 percent of the energy produced in the United States to come from renewable sources by 2025.

Provide long-term assurances of investment and production tax credits. Depending on how you calculate things, solar power and wind power are still years or even a decade from being fully cost competitive, but we don't have that kind of time to kill. And when it comes to spurring action, the evidence is overwhelming. In the years Congress has passed tax incentive legislation, investments in renewable energy sources have skyrocketed. In the years those incentives failed on Capitol Hill, those investments took a nosedive. Pay-as-you-go conservatives should not be allowed to block these incentives with their false economic analyses that ignore the long-term savings to be gained by earlier adoption of renewable technologies. Moreover, these incentives should be designed to last at least a decade, to provide the kind of stability and assurance that will prove truly inviting to investors.

Revive federal investment in renewable energy research and development. Time was, a few decades ago, that the Department of Energy (DoE) invested billions annually in renewable energy R&D. Indeed, the federal government accounted for 98 percent of solar research in the country. Today, DoE's renewable energy budget is a mere $250 million or so. And most of that is in photovoltaics, even though the other major method of capturing solar energy—thermal capture, which grabs heat from the sun and uses it to power steam turbines—is equally deserving. Today DoE's funding of solar energy R & D is just 2 percent of the national total, a pittance given the importance of government investment at this crucial stage of technology development.

Of course, making the transition to renewable energy will not be simple. Environmental issues will have to be addressed, including the effects of large-array solar facilities on endangered species such as the desert tortoise. Companies have also complained, with some legitimacy, that as empty as the Southwest deserts seem, the confusing patchwork of federal and private land ownership and land-use restrictions combine to pose, in the words of Steve Kline, a vice president at

Pacific Gas & Electric Co., "a challenging permitting process" in the California desert.

There also remain lingering technological challenges. Solar thermal capture systems are virtually 100 percent efficient already, but construction costs could still be lowered considerably. Photovoltaics must become more efficient and cost-effective—and are expected to do so in double-digit leaps over the next few years as better physics and chemistry are brought online, especially through nanotechnology. Energy storage must be improved, since sunlight is obviously no way to power a city at night. But studies show that energy demand actually tracks peak sunlight times quite closely, so an ability to store captured energy even for just an hour or two would go most of the way toward allowing full reliance on sunlight.

Finally, the nation's infrastructure of transmission lines needs to be upgraded, through a creative public-private partnership, to carry energy cheaply and efficiently from the places sun and wind are most prevalent to regions that are lacking. That is an interstate-highway-like project that the new president should take on quickly and with Eisenhower-ish zeal.

By one estimate, the mere elimination of scattered, small, but constantly occurring power interruptions caused by storms and other insults to our aging transmission lines would save something like $80 billion a year. Savings like those could offset upgrade expenses considerably. But for those looking for a source of cold cash up front, to prime the renewable energy research pump, there's plenty of that to be had as well—in the suitcases of money now being spent on federal tax breaks and subsidies for the oil and gas industries.

ExxonMobil Corp. reported third quarter profits of $14.8 billion in the autumn of 2008, the largest in history and almost 60 percent higher than a year earlier. Ongoing subsidies for a company as well heeled as that are about as anachronistic as a satellite that can do no more than beep.

Owning Up to
Our Oceans

VERY SUMMER, LIKE CLOCKWORK, A VAST BODY OF NUTRIENT-CHOKED
and oxygen-depleted water, roughly the size of New Jersey, forms
off the mouth of the Mississippi River in the northern Gulf of Mexico.
Fed by millions of tons of nitrogen- and phosphorus-rich agricultural
runoff, this hypoxic region—or, as it's more commonly known, dead
zone—has been expanding at an alarming rate over the past few
decades as fertilizer and fossil-fuel use have surged.

Along with a separate but equally worrisome problem—ocean
acidification, caused by atmospheric buildups of carbon dioxide—the
growth in dead zones around the world is an ominous sign that human
environmental impacts have achieved a scale once thought impossible.
Having soiled the soil and blackened the air, we are now tipping the
balance of Earth's largest and most stable ecosystem, the planet's seemingly imperturbable oceans.

Of all the dead zones that now pock the planet's oceans, the gulf's
may be the best studied and most infamous, situated as it is at the
mouth of America's largest river. But many others—more than 43, at
last count—have sprung up around the United States in recent
decades, most noticeably in the Chesapeake Bay and off the coasts of
Oregon and Washington. And of course this is not merely a domestic
concern. Ranging widely in size from small areas in coastal bays to vast
swathes of water in the open ocean, dead zones exist around the globe,
most of them in temperate seas and primarily around highly developed
countries. According to a United Nations report, there are now more
than 150 documented dead zones around the world. Studies have

found that the number of dead zones has roughly doubled every decade since the 1960s.

And here's the kicker: while most are still seasonal, climate change could prolong these events—and make them much more frequent.

One of the world's longest river systems, the Mississippi River drains 41 percent of the contiguous United States. At the mouth of this system lies the planet's second-largest zone of oxygen-depleted waters—the gulf dead zone—which in 2008 had an estimated area of more than 10,000 square miles. Over the past few decades, a number of studies have cast light on the causes and impacts of such hypoxic events.

Perhaps not surprisingly, the enemy, in large part, is us. Specifically, rising farm production and energy consumption have led to greater runoff of fertilizers and other compounds, causing imbalances in the global nitrogen and phosphorus cycles. Increased concentrations of these key nutrients in river effluents can cause a rapid growth of phytoplankton, microscopic plantlike organisms, producing massive blooms in a process known as eutrophication. In time, as these carbon-rich morsels grow old and deteriorate and sink to the seafloor, they are feasted upon by a vast array of microorganisms, whose respiration consumes all of the available oxygen in the surrounding waters. That creates anoxic, or dead, zones.

How dead? Well-oxygenated waters typically contain up to 10 milligrams of oxygen per liter, or 10 parts per million (ppm). By contrast, the concentration of dissolved oxygen in hypoxic or dead zones often hovers below 2 ppm. In some cases, it can plunge below 0.5 ppm and remain there for several months. The result is not just a decline in biodiversity, which would be worrisome enough. In many cases we're talking about an area completely devoid of life.

Things appear poised to get worse. U.S. farms are expected to produce record amounts of heavily fertilized food crops in coming years. And the federal government is signaling that it may free up currently protected land to plant additional acres of corn, to help meet the demand for ethanol as a renewable fuel. Moreover, global warming may already be aggravating dead-zone events off the coasts of Oregon and Washington. Marine ecologist Jane Lubchenco, the new head of the National Oceanic and Atmospheric Administration, suspects that the stronger offshore winds produced as land heats up are prolonging upwelling in coastal waters. Upwelling is the process by which deep, nutrient-rich waters are drawn up to the surface by winds; it provides a vital source of food that stimulates much of the ocean ecosystem's primary production. In this case, however, too much of a good thing can be harmful. An excess of phytoplankton that isn't consumed will die and fall to the seafloor, where its decomposition can create large, oxygen-free zones. Worse, Lubchenco and her colleagues have found that these low-oxygen areas, which typically reside in deep waters, are spreading to shallow fishing waters—a discovery Francis Chan, a fellow ecologist, has described as "unprecedented."

Efforts begun by the federal and state governments in 2001 to rein in these problems have produced little in the way of results. A coordinated federal plan to shrink the dead zones by making cuts to nutrient runoff never made it past the budget process once the Bush administration took office. A revised plan led by the Environmental Protection Agency has the dual objectives of shrinking the gulf dead zone to about one-quarter of the size of the summer 2007 zone by 2015, and of slashing nitrogen and phosphorus levels by 45 percent each by that year. But the plan gives states until 2013 to complete their implementation strategies, leaving as little as just two years to achieve the necessary reductions. That has struck some scientists as an unrealistic and ultimately toothless policy.

At this rate, it is clear that we may be close to reaching a tipping point after which dead zones will be considered the "new normal," as Lubchenco puts it. The consequences will be devastating: completely altered ecosystems, dwindling biodiversity, and exhausted fisheries populations, to name a few. And all of it likely to be exacerbated by the second major insult ongoing in our oceans: acidification, caused by the unprecedented influx of anthropogenic carbon dioxide emissions since the 1800s.

Scientists have long been aware of the essential role played by the ocean in mitigating the impact of elevated atmospheric carbon dioxide (CO_2) levels, thanks to the groundbreaking research of Roger Revelle and Charles Keeling dating back to the 1950s. Ice core measurements of carbon dioxide taken midcentury showed that atmospheric concentrations had remained about constant for several thousand years until the rapid onset of industrialization during the 1800s, after which they began their meteoric rise. Revelle's work was instrumental in demonstrating that although a large fraction of the gas remained in the atmosphere, a significant amount was also being absorbed by the ocean—a realization that would lead him to conclude that, over the long term, it would permanently change the chemistry of seawater. A number of oceanographer-led global surveys completed in 2004 determined that the ocean had, in fact, absorbed nearly half of all carbon emitted since the start of the Industrial Revolution. Other studies have found that around a third of fossil-fuel-derived CO_2 is currently being taken up by the ocean.

Upon entering the ocean, a portion of CO_2 reacts with water to form carbonic acid, a weak acid; the other portion stays in dissolved form. Some fraction of the acid releases hydrogen ions into solution, yielding either bicarbonate or carbonate ions, while a smaller fraction remains as carbonic acid. These three forms of dissolved inorganic carbon—carbon dioxide, bicarbonate ions, and carbonate ions—together make up a natural buffer, called the carbonate buffer, which helps stabilize ocean acidity by absorbing small pH changes induced by the increase in hydrogen ion concentration. The pH scale, which ranges from zero to fourteen, is used by scientists to measure a solution's acidity or alkalinity—the lower the value, the more acidic the solution. The scale is logarithmic, so a one-pH-unit drop corresponds to a tenfold increase in the hydrogen ion concentration, making seawater considerably more acidic. With an average pH of 8.1, seawater is normally slightly basic, or alkaline.

This buffering system has helped keep the ocean's pH in check for thousands of years. However, the unprecedented influx of anthropogenic CO_2 emissions since the 1800s has fundamentally altered the equation, threatening to overwhelm the delicate balance maintained by this system and tipping the ocean into a period of prolonged acidification. The problem is simple: as increasing amounts of atmospheric CO_2 are absorbed by surface waters, more hydrogen ions are formed—which leads to an overall decrease in seawater pH. Many of these hydrogen ions will combine with carbonate ions, forming bicarbonate ions and reducing the concentration of carbonate ions. The net effect is to weaken the carbonate buffer, rendering it less effective at keeping slight pH variations in check.

The magnitude of this change may sound small, but can be chemically and biologically very significant. Researchers believe this process has already lowered the ocean's average pH by 0.1 since the preindustrial era—equivalent to a 30 percent increase in the ocean's average hydrogen ion concentration. A recent analysis postulated that pH levels might fall by as much as 0.5 units by 2100, which would be equivalent to a threefold increase in the hydrogen ion concentration since preindustrial times.

What difference can this make? For one thing, by reducing the ocean's ability to serve as a carbon sink, more carbon dioxide will have nowhere to go but into the atmosphere, contributing further to the widely recognized climate problems related to the greenhouse effect. But aside from that, what worries scientists most about ocean acidification is that it will inhibit certain organisms' ability to produce calcium carbonate shells, which is normally accomplished by means of a biochemical precipitation reaction that depends on a well-balanced carbonate buffer system. Especially at risk are tiny phytoplankton, which support entire marine food webs by acting as the ocean's primary producers, equivalent to plants in terrestrial ecosystems. Without them, or with their numbers greatly reduced, many populations and ecosystems could simply collapse.

Ocean life, in some form, would go on, of course, and some species would even likely benefit. A few recent studies have demonstrated that some phytoplankton species may thrive in elevated CO_2 concentrations, and some larger organisms, such as sea grasses, which use dissolved carbon dioxide directly, could also experience gains. But the losses, over all, would be enormous, and would likely include corals, one of Earth's watery wonders. Corals are not just pretty tourist attractions but constitute critically important habitats for large numbers of marine species. They are tiny animals that secrete calcium carbonate skeletons around themselves, which, over time, accumulate to form large reef assemblages. Alas, in recent years many of the world's reefs have already succumbed to disabling and even fatal "bleaching" episodes, apparently as a result of changing ocean chemistry. The algae that live in symbiotic associations with corals, and that give corals their colors, are expelled in this process, depriving the corals of a critical source of energy and nutrients.

By some estimates, all of the planet's corals could disappear by century's end if present trends continue. That would be a devastating legacy for today's planetary stewards to leave the next generation.

Going with the Flow

ONE OF THE MOST DISTINCTIVE LEGACIES OF THE NINETEENTH-CENTURY progressive era in the United States was the ambitious and successful effort to harness the mighty western rivers. Today, nearly every river in the West is regulated by dams, locks, or diversions to provide water for irrigation and electricity for growing cities and industries.

These dams and their extensive water-distribution facilities fueled an engine of growth and prosperity that drew millions to the region. The dams on the Colorado River alone are capable of storing four years' worth of river flow. Unfortunately, the very infrastructure that today makes the desert bloom has nearly destroyed that river's native fishery and has fundamentally altered and undermined the ecosystem it supports.

The same is true across the region. Damaged wetlands, endangered fish and wildlife, and impaired communities and economies are commonplace today where healthy and intact river systems once existed. But the fact that our monumental western water infrastructure has hurt the environment is old news. The newer and better news is the vast amount of work that is now underway to restore the region's rivers and their associated natural and human communities.

We are, in fact, already well embarked upon a twenty-first-century progressive era in the American West—one in which the federal agencies once known for the cubic yards of concrete they poured are now directing increasingly significant resources to restore rivers, wetlands, and riparian corridors. This is an encouraging movement, and one worth celebrating and strengthening.

Take the U.S. Bureau of Reclamation, which got its start in 1902 in the same burst of legislation that gave rise to the Newlands Irrigation Project. That project diverted water from the Truckee River near the California-Nevada border to irrigate high-elevation desert farms. Under-appreciated at the time, the water flowing to Newlands depleted the river's terminus, Pyramid Lake. Eventually, lake levels fell by seventy-five to eighty feet, nearly wiping out the native fish populations and thus preventing the Pyramid Lake Paiutes, whose reservation sur-rounds the lake, from exercising their historical fishing rights.

Decades later, following lawsuits and lengthy negotiations, Congress enacted in 1990 the Truckee-Carson Settlement Act, which directed the Bureau of Reclamation and other parties—including the states of California and Nevada, other federal agencies, the Paiute tribe, and private water interests—to find new ways to work together to restore the river, the lake, and the fisheries. This has proved to be a challenging mandate, but today an impressive multiparty restoration initiative is well underway.

The manager of this project for the Bureau of Reclamation, Elizabeth Rieke, previously was the assistant secretary of water and science for the U.S. Department of the Interior. She possesses a keen grasp of fed-eral water policy. Commenting on the changing mandates for federal water projects at a conference sponsored by the Buffalo Bill Historical Center, Rieke remarked, "We can build them, operate them, modify them, re-operate them, we can make them safe and secure, and we can take them down."

Her message, in short, was that the same technical expertise that erected the West's great water projects can be harnessed in new ways to benefit a broader range of public values.

For its part, the U.S. Army Corps of Engineers—the agency respon-sible for constructing the largest flood-control dams in the nation—has begun to embrace environmental protection and restoration as explicit objectives of its water resource management mission. In one example of its commitment to this new direction, the corps has entered into a sustainable rivers partnership with The Nature Conservancy, aimed at improving dam management to restore ecological health in the affected rivers.

River restoration projects are proliferating throughout the Rocky Mountain West. In some cases this means that dams are coming out of the rivers they once plugged. More commonly, federal agencies are operating dams in new ways to re-create historical downstream river conditions. In many places, restoration means putting the curves back into artificially straightened rivers, replacing riprap (stone or rubble dumped along river shores to combat erosion) with native vegetation to secure the banks, and exposing rivers once buried in steel culverts under urban centers—a practice known as daylighting.

Recently, onlookers cheered as Montana Gov. Brian Schweitzer declared, "Let 'er run," and a large bulldozer breached the Milltown Dam, a few miles upstream from Missoula. Today, the waters of the Clark Fork and Blackfoot rivers run free for the first time in 100 years.

Another measure of success: a large Superfund cleanup project at the rivers' confluence is underway to remove tons of contaminated sediment that flowed downstream from copper mines in Butte and stacked up behind the Milltown Dam, polluting the drinking water of nearby residents. The costs of such cleanups can be high, but the benefits can be enormous. And the restoration work itself provides jobs for skilled laborers and professionals. The Milltown Dam removal has employed close to 100 people, including heavy-equipment operators, engineers, and scientists.

Like the reclamation projects of the last century, today's restoration initiatives represent an investment in the future, with long-term benefits for the environment and its human inhabitants. By recognizing the value of this work and encouraging it through more explicit public policies, we can ensure a healthier and more prosperous future for all. That is the heart of a new progressive movement for the Rocky Mountain West.

But there is another side to getting it right with water in the West: slowing the stampede of development in areas where water supplies— even well managed ones—simply can't support ongoing growth. It may seem obvious that it makes no sense to approve development without reliable water supplies. But that is exactly what has been happening for many years in communities throughout the American West—leaving homeowners and other taxpayers holding the bill when extravagant measures become necessary to gain access to water.

It's a disconnect not tolerated in other arenas of development. Homeowners expect safe wiring and solid foundations for their dwellings, and building codes demand it. In the same way, residents in arid communities deserve to know that their taps will deliver clean, reliable water for decades to come. Yet historically, land-use decisions and water planning have been treated as entirely separate issues. Water is allocated by state agencies, while land-use planning falls under the authority of local officials. Water resource managers juggle many competing demands within a watershed, and they tend to focus on facilitating economic development. In turn, local land-use authorities have largely assumed that water would be available to satisfy continued growth. And until recently they could generally get away with that gamble.

Increasingly, however, local land-use decisions have begun to run headlong into water-supply concerns. In some cases, existing uses are depleting finite water supplies, raising questions about their future reliability. In some fast-growing rural areas of Arizona, for example, recently constructed houses are drawing water from wells that the state engineer's office has certified as "not reliable" due to insufficient underground supplies. Some new homeowners did not realize the tenuous nature of their water supplies and have been forced to deepen their wells or construct cisterns and pay for trucked-in water.

Elsewhere, officials are beginning to face the high social, environmental, and economic costs of obtaining water to meet rising urban demands. Urban growth around Phoenix, Denver, and Boise has been fueled by voluntary, market-based reallocation of water from farms to cities. But public outcry over Las Vegas's long reach into rural Nevada is evidence of renewed concerns over the impacts of large-scale water transfers, both on the rural communities from which the water is taken and on the pocketbooks of the consumers receiving it.

Happily, as is happening with western river restorations, progressive change appears to be gaining ground. In Colorado, for example, the General Assembly recently took an important first step toward ensuring reliable water supplies for new development: it passed a law that helps local governments determine the adequacy of water supply for proposed development projects. It also gives local governments the discretionary autority to deny developments if adequate water supplies cannot be assured.

The Colorado bill does not assign any time horizon to the supply requirement but simply looks at the possible peak daily, monthly, and yearly demands at projected build-out levels of development. Other states, including Arizona and California, now go further, requiring "assured supplies" for 50- to 100-year planning horizons, although each state has significant exceptions built into those requirements.

What would an ideal assured-supply law look like? According to Utah law professor Lincoln Davies, such a law would be mandatory, stringent, statewide, broadly applicable to large and small projects, and interconnected with broader planning mechanisms for land, water, and environmental protection. Thus far, no state statute meets all these criteria, though legislation enacted in California comes close.

Acting under the mandate of that legislation, last year the California Supreme Court said "No" to a planned mixed-use development in the Sacramento area on the grounds that the long-term water-supply plan was inadequate. Similarly, the Riverside County Superior Court also recently blocked a large planned development because of water-supply concerns.

Such actions come with a price, of course. In California, state water managers and Gov. Arnold Schwarzenegger have expressed concerns that unreliable water supplies and resultant development delays may be destabilizing the state's powerful economy. The San Diego civil grand jury went further in a recent report with the attention-grabbing title "Sober Up, San Diego. The Water Party Is Over," which concluded that permanent, mandatory conservation measures would be necessary if the arid landscape is to continue supporting vast numbers of people.

Such strong public statements remain the exception rather than the rule, but the trend is clearly toward taking a harder look at water supplies before approving new development. Western expansion has long relied on the promise of abundant and cheap water—a myth that is already shattered in many communities and is sure to be exposed as false in many more in the decades to come.

Pruning Old
Promethean
Policies

WILDFIRES CHARRED MORE THAN 4.5 MILLION ACRES NATIONALLY IN 2008. That's a lot of charcoal, but for those in the know it is almost good news. In each of the four previous years, fires swept over more than 6 million acres, and the ten-year average leading up to 2008 was more than 5.6 million.

Indeed, 2008 proved to be somewhat of an aberration—a fairly mild fire year across a Rocky Mountain West, where massive, sometimes uncontrollable fires have become commonplace in recent decades. Happily, the winter of 2007–2008 brought ample snow across the mountainous West, and helpful rains fell through the spring and summer. All told, abundant moisture helped limit the size and severity of fires.

But those conditions were far from typical and should not give rise to complacency among those concerned about the ecological health of the American West. Studies indicate that the frequency and severity of wildfires have increased dramatically in recent decades, and there is no reason to believe that the trend will reverse itself as global climate change runs its course.

The historical perspective is enlightening. Warmer springs and longer summer dry periods since the mid-1980s have helped to cause a fourfold increase in the number of major wildfires each year and a six-fold increase in the area of forest burned, compared with the period from 1970 to 1986, according to the National Wildlife Federation. Today's fire season is seventy-eight days longer than it was a generation ago.

Statistics compiled by the National Interagency Fire Center in Boise, Idaho, illustrate the trend well: in 1997, fires burned some 2 million acres in all. In 2007, the total was more than 9 million acres. As the National Fire Protection Association's Jim Smalley said, the potential for "megafires" is increasing annually, and nearly every year to come promises to become "the worst fire season ever." Although 2008 proved to be an exception, the unmistakable long-term trend is toward a warmer, drier West. Climate scientists project global climate change will bring summertime temperatures in the region that run 3.6 to 9 degrees Fahrenheit higher than what's now considered normal. Those scientists also predict that precipitation will decline as much as 15 percent by the middle of this century. The result will be a climate even more conducive to wildfire.

Of course, we've always had wildfires across the West. Over the past 10,000 to15,000 years, fires sparked by lightning have shaped our forests, and over the past few centuries human-ignited fires, too, have had a significant impact on natural systems. Some (mostly lower elevation) types of forests evolved in an ecological context of frequent but low-intensity wildfires. Other forest types have evolved to accommodate more intense but less frequent fires. Indeed, the natural role of fire is generally beneficial—clearing out underbrush, holding disease and insect infestations in check, and aiding the regeneration of some types of trees.

Humans, however, have interrupted and exacerbated natural fire cycles. After massive wildfires swept through the Northern Rockies in 1910, federal and state land-management agencies adopted a policy of aggressive fire suppression with a goal of quickly extinguishing fires wherever and whenever they occurred. And they were remarkably successful. The U.S. Forest Service, for example, maintains better than a 90 percent record of stopping fires when they're small. But as ecology and fire historian Dr. Stephen Pyne explains in *Forest History Today*, a publication of the Durham, N.C.–based Forest History Society, "withholding fire is as powerful an ecological act as applying it."

In recent decades, it's become clear that interrupting natural fire cycles has allowed forests to grow unnaturally dense without periodic fires to clear out underbrush and smaller trees. Lack of fires has allowed species of trees not adapted to frequent fires to outcompete thicker-bark tree species that thrive with frequent fires. Without fires, smaller trees now create "fuel ladders" to carry flames into the crowns of larger trees that otherwise might not be scathed by fires burning along the ground.

Moreover, without fires, insect infestations have begun to reach epidemic levels, creating forests of dead, standing trees. Colorado, for example, has some 1.5 million acres of pine beetle–killed forests primed for burning. In Montana, a fir beetle epidemic more extensive than any other documented in history has killed Douglas firs over hundreds of thousands of acres. Across the West, forests have become less healthy and productive—with far, far more potential for larger and often unnaturally intense wildfires.

As the Government Accountability Office put it in a 1999 analysis, "The most extensive and serious problem related to the health of national forests in the interior West is the over-accumulation of vegetation, which has caused an increasing number of large, intense, uncontrollable and catastrophically destructive wildfires."

Compounding matters, the forest fringe remains an attractive place for people to live. The Rocky Mountain West has become one of the fastest-growing parts of the country, and much of that growth has taken place in the form of residential development in proximity to our now-overgrown, fire-prone forests.

This combination of factors—a warming climate, unhealthy forests, and rampant development—sets the stage for wildfires of great size, intensity, and danger.

The potential for large fires places tens of thousands of homes and entire communities in harm's way. Protecting lives and property has become hugely expensive for local governments, states, and the federal government. Annual federal firefighting expenses have risen more than sixfold over the past decade and now exceed $1 billion in a typical fire year—nearly $2 billion in 2007—with more than half that money spent to protect homes and communities, a form of mission creep for agencies primarily responsible for land management. The U.S. Forest Service now expends nearly half its annual budget on wildfire suppression at the expense of other needed forest management, including measures that could reduce fire danger over the long term.

The West's growing fire danger defies simple solutions. One thing is clear, however: the health of our forests—and our own health, safety, and prosperity—depends on managing fire rather than strictly suppressing it. In forest ecosystems that evolved with frequent, low-intensity fires, for example, land managers can use prescribed or controlled burning to duplicate natural cycles in a relatively safe way. In a report commissioned by Western Progress, an independent, non-profit policy organization with bases in Phoenix, Denver, and Missoula, W. Wallace Covington and Diane J. Vosick of Northern Arizona University's Ecological Restoration Institute make a compelling case for actively managing fire danger and restoring forest health and productivity through restoration forestry.

We'll have difficulty managing fire—learning to live with fire—as long as the buildup of forest fuels around houses, neighborhoods, and communities leaves too much potential for the loss of property and lives. Before we can make greater use of controlled fires as a management tool or allow more wildfires to burn naturally, we have to do something about the condition of today's forests. We have to reduce the amount of accumulated fuels. Estimated costs of doing so vary, but $750 to $1,000 an acre is a good ballpark figure. Whatever the cost, it's massively expensive work to do over the 73 million acres of national forests and 397 million acres in total deemed a high priority for fuel reduction. What's more, fuel reduction involves tree thinning, which is logging, and that remains a highly contentious issue in the political arena as well as in the nation's courtrooms, at least whenever national forests and other public lands are involved.

But broad consensus does exist among land managers and the public over the merits of forest-fuel reduction in proximity to homes, neighborhoods, and communities—an area best known by its bureaucratic moniker: the wildland-urban interface. Aggressive fuel reduction in the wildland-urban interface will make people and their property safer. It won't fireproof neighborhoods, but it will make fires less threatening when they reach thinned-out areas. It will make firefighting safer and more effective, generally resulting in less intense fires approaching homes and communities. And it will give land managers more flexibility as they work to reduce fuels over a broader swath extending into the national forests. For example, it should be safer and easier to use controlled fire elsewhere in a national forest if thinning has removed excess fuels near communities. As a fringe benefit, fuel-reduction work over the millions of acres that make up the wildland-urban interface in the West has the potential to employ thousands of workers in rural communities across the West as part of the region's emerging restoration economy.

The value of fuel reduction in proximity to homes and communities is well understood by land-management agencies and, increasingly, by the public. But funding for such work is grossly inadequate—and often diverted to pay for fire suppression. That rob-Peter-to-pay-Paul approach has made it impossible for federal and state agencies to proactively reduce forest fuels on a large scale. As a result, taxpayers spend increasing sums reacting to growing wildfires. With current western burn rates, alongside the sluggish pace of fuel reduction, the region's forests will burn over before thinning work gets done. What's needed—from Congress and state legislatures—are new streams of dedicated funding that can ensure steady progress toward meaningfully managing the region's fire risk.

HARNESSING HEALTH, BOLSTERING BIOMEDICINE

No field of knowledge has progressed more quickly in recent decades than that of biomedicine and genetics. A little more than a half century since James Watson and Francis Crick first deduced the three-dimensional structure of DNA, scientists have identified and placed in order all 3 billion "letters" of the double helical code that together spell out the directions for making and maintaining a human being. And for better or worse, having gained experience splicing those pieces in and out of bacteria, worms, plants, and farm animals, we stand on the brink of an era in which our personal genetic fates are becoming predictable and perhaps even subject to engineered solutions.

At the same time, on the level of the cell, scientists have come to a remarkable degree of understanding of the processes that underlie both the ravages of cancer and the promise of regenerative medicine, suggesting that the already astonishing increase in the average American life span from just forty-seven years in 1900 to more than seventy-eight years today may continue apace and soon make centenarians as common as eighty-year-olds are today. As the following writings suggest, what we do with that power and how we share its benefits with those who might not on their own have access to it will say a lot about what we are as a civilization.

Prescription for a Bodily Extended Warranty

Sometimes people get a bad reputation just for being a little ahead of their time. Consider the explorer Juan Ponce de Leon. For having sought a fountain of youth, the guy earned a place in history as a frivolous narcissist. But was he really so off the mark? An amazing string of discoveries in recent years suggests that the concept of slowing and perhaps even reversing aspects of biological aging is not so ridiculous after all. Indeed, a growing cadre of scientists is today beating the drum for a significant shift in the way medical-research dollars are spent. The idea is that researchers can accomplish more by learning how to slow the aging process than they can by making incremental gains against individual diseases of aging, such as cancer and heart disease.

Life expectancy has grown by leaps and bounds in the past century. And to be sure, that success story has been written not by exotic antiaging creams or wrinkle removers but by hard-won victories over infectious diseases, primarily through public-health advances and antibiotics. A century ago, only about 40 percent of babies born in countries where births and deaths were reliably counted could be expected to live past age sixty-five. Today, almost 90 percent do so. That's one reason why in the United States alone, the number of people older than eighty-five is projected to quadruple by 2050, to 18 million from today's 4 million.

But how do we keep that progress going in developed countries, now that infectious diseases are no longer the major killers in those areas? The current approach to extending life, which has garnered some modest success in the past decade or so, has been to focus on curing those chronic diseases of aging that have become the modern era's major killers, such as heart disease, stroke, cancer, diabetes, and Alzheimer's disease.

Yet most of these diseases have proven remarkably resistant to treatment. More important, a number of research models have concluded that even if scientists were to score a complete home run by finding a cure for any single chronic disease such as cancer or stroke, life spans in developing countries would hardly grow longer. That's because the other chronic diseases of old age are right there, waiting to kill us anyway at about the same age. You'd have to find cures for virtually all of them, these models suggest, to make any real progress at this point.

One consequence of this disease-by-disease approach has been that a lot of elderly people today are alive but chronically ill. That is taxing not only public -health-care resources but also these patients' personal savings and those of their family members, who in many cases are saddled with a large part of these patients' burden of care. In fact, 80 percent of the nation's long-term care is provided by unpaid caregivers, mostly family and friends. And while there are many obvious benefits of keeping at least a portion of long-term care in the family, it is not practical in many cases to expect family members to be able to carry the entire burden. At least 60 percent of those unpaid caregivers are already busy with their own jobs.

Meanwhile, few employers offer assurances of help for employees who need to care for aging relatives. And the Family and Medical Leave Act, which assures up to twelve weeks of unpaid leave for such purposes, still applies to only a fraction of employers and, if anything, has lately come under threat of being weakened by business lobbyists.

These social and economic costs add credence to the new, emerging approach to diseases of old age: the across-the-board antiaging approach. As it turns out, scientists studying the aging process in a wide variety of organisms—from worms to flies to fish to monkeys — have been finding that certain genes common to virtually all kinds of life play key roles in the metabolic and genetic processes that together underlie most diseases of aging. Some of these genes are important for repairing DNA damage caused by sunlight, chemicals, or other environmental insults. Others affect the levels of vital hormones. As scientists figure out what these key genes do, it becomes possible to envision developing medicines that mimic their activity, which typically wanes late in life.

Encouragingly, this work is not limited to nonhuman species. A number of ongoing research projects have recently begun to clarify some of the most important antiaging genes in people. In one project, researchers are comparing DNA from two populations: people eighty or older who have never had serious illnesses and those who died of age-related ailments before they hit eighty. The project aims to find DNA coding that helps protect some people from the standard ravages of old age.

Unlike conventional advances in life span, which in many cases have added years of disability and suffering to the end of life, the antiaging approach would add healthy years, according to Colin Farrelly of the University of Waterloo in Canada. "There is a credible scientific basis for believing that we could slow aging in the foreseeable future," he concludes. "And the amount of public funding we invest into such research will determine the likelihood and timescale of success for aging interventions."

Some proponents, including Robert Butler (a former head of the federal National Institute on Aging), S. Jay Olshansky (a leader in the antiaging field at the University of Chicago), and Daniel Perry (of the Alliance for Aging Research in Washington), argue that $3 billion, or just 1 percent of the Medicare budget, would be a "prudent" investment in antiaging research. They make a convincing case that such an investment could pay for itself many times over by delaying not only the fatal diseases of old age but also the many debilitating conditions such as osteoporosis, arthritis, cataracts, and cognitive decline that typically take a big toll on quality of life in people's final decade or two.

They call that benefit "a longevity dividend," and it promises a much higher rate of return than most of the retirement products you'll find on the market. Meanwhile, hang in there, ol' Ponce. Your reputation, if not your bod, may get rehabilitated yet.

Mom, Apple Pie, and Interoperable Electronic Medical Records

THEY DON'T TEACH IT IN MEDICAL SCHOOL, BUT PEDIATRICIANS KNOW that the most reliable and effective repository for medical information has for decades been a mother's purse. Through changes of employment, moves to new towns, and seasonal upgrades of the purses themselves, children's immunization records always made the move along the journey from infancy through college.

Every mother keeping track of those immunization cards understood the importance of having them available for the next doctor or emergency room visit. The records' availability prevented her child from incurring unnecessary pain and possible vaccine reactions. In the same way, mothers of chronically ill children have long been the trusted couriers of important office notes and lab tests ordered by the multiple specialists who saw their children.

This informal information exchange has never been meant as a substitute for the written physician-to-physician consultations that are included in conventional medical records. But they have facilitated communication, especially as the number of specialists has increased. And mothers almost never lose records.

In modern lingo, generations of mothers have instinctively defined the need for multiple user capability, or interoperability, the technical term for the primary functionality in electronic medical records. Medical professionals recognized the potential for EMRs as far back as the 1970s, but with recent rapid advances in Internet technology, digital records today can truly revolutionize the way we do medicine in the

United States, at last matching and even surpassing the benefits long embodied by the old "maternal medical records" system.

The current paper-chase method requires a mother to define what modern experts would call "multiple user input and access functionalities" to accomplish what she already knows is important—timely availability of medical history to avoid duplication of medical interventions and their unnecessary pain, complications, and costs. A real-time interoperable EMR system eliminates those racks of paper records lining doctors' offices and mothers' purses across the country. But more importantly it would improve the quality and affordability of care from individual physicians and ensure that what goes on in one doctor's office helps the next doctor's office provide better and more affordable care.

According to a study in the *Annals of Internal Medicine*, physicians with paper records could perform as well as physicians with an EMR system, but that study only looked at the use of an electronic record within the "silo" of a single physician's office. It did not look at the efficiency that interoperability could provide by breaking down the barriers to communication among different doctors' offices.

The promise of interoperability is its ability to integrate electronic medical records across the health-care system, from hospital to lab to primary-care physician to medical specialist. This integration allows each health-care provider in the chain to have accurate, up-to-date information with which to make decisions for the next step in care— and in a timely fashion without duplication or indecision on account of not knowing what happened in another office.

What's more, a key feature of an interoperable electronic medical record is that it allows patients (or their mothers) to define which caregivers would get electronic access to medical information across the health system. Patients thus have control over who sees their EMRs, which, in tandem with current rules under the Health Insurance Portability and Accountability Act of 1996 governing the disclosure of private health information by medical-care providers and commercial payers, goes a long way toward addressing privacy concerns about EMRs. Further steps need to be taken to ensure complete privacy protection, but these concerns should not get in the way of a much-needed EMR rollout.

Still, no digital system can ensure that health-care providers use the information at hand unless a critical mass of providers embraces its use. Unless EMR systems are tapped for information, even the best interoperable system cannot guarantee a quality result.

A recent study in the *Annals of Family Medicine* discovered just that pitfall. The study, which focused on the use of EMRs in diabetes care, concluded that:

> The use of an EMR in primary care practices is insufficient for insuring high-quality diabetes care. Efforts to expand EMR use should focus not only on improving technology but also on developing methods for implementing and integrating this technology into practice reality.

That's why the federal government needs to take the lead in making EMRs a reality in doctors' offices across the country.

How do we accomplish this? Past changes to the health-care system demonstrate that when the federal government wields its power as the largest health-care payer, commercial insurers move to adopt the government standards. Case in point: Medicare adopted the diagnosis-related group classification system to define the primary reason for a patient's hospital admission and determine payment. Now virtually all physicians use the system. Or consider the Food and Drug Administration's guidelines for coverage of new medical-device technologies. When the FDA adopted the guidelines thirty years ago, health plans began relying on approval as the first step in determining whether commercial insurers would consider paying for the new device. In each instance, the federal government implemented a process that improved the health-care system, and commercial insurers followed suit. We could expect the same kind of large-scale buy-in with interoperable EMRs if federal agencies take the lead.

The sooner the federal government mandates and funds an interoperable EMR platform, the sooner our health-care system will start to mend in a cohesive fashion. The incentive for the federal government to do this is the enormous amount of time and money the Medicare system wastes year after year because practice performance measurements of guideline-driven care are not implemented for patients with chronic illnesses.

According to a Rand study published in the *New England Journal of Medicine*, only 55 percent of patients received care according to professionally agreed, upon guidelines. Whether the care is too little too late or too much too often, resources are wasted and patient outcomes fall short. An interoperable EMR can identify where guidelines aren't being met for those patients being treated by multiple providers during the course of their illness. Each provider would have the information and could potentially improve care where guidelines are not being followed. The patient could receive all the care needed, and none that he or she didn't need.

It would make your mother proud, and it could put more than just an immunization card back in her purse.

Your DNA's
Spitting Image

JEFFREY GULCHER HAD NO REASON TO THINK MUCH ABOUT PROSTATE cancer. He was just 48, and the disease typically strikes relatively late in life. Even the most cautious medical groups agree that most men need not begin annual prostate screenings until age fifty.

But Gulcher happens to be the chief scientific officer of deCODE Genetics, one of several companies that, amid some controversy, have begun offering direct-to-consumer DNA tests that can help people predict which diseases they are likely to get. So in April 2008 he spit into a test tube and, without giving the matter much thought, sent the sample in for analysis by his own company.

He was in for a shock. The test indicated that he carries a genetic variant that nearly doubles his lifetime risk of getting prostate cancer: while the average man has a 15 percent chance of being stricken, Gulcher had a 30 percent shot. That spurred his physician to order a standard blood test for prostate cancer. The result was within the range considered normal. But it was toward the high end of that range, which, together with the initial DNA test, worried the doctor. He referred Gulcher to a urologist, who performed an exploratory biopsy—and found that Gulcher's prostate gland was riddled with cancer, and a fairly aggressive version of it at that.

Not long after he got his results, Gulcher went in for surgery, and not a moment too soon. The disease had not yet spread to other parts of his body, a milestone that often portends death and that may well have been passed had he waited until he turned fifty to get a standard prostate-specific antigen (PSA) test.

Did genetic testing save Gulcher's life? It may well have. Indeed, his dramatic story seems to illustrate perfectly the claims, made by his company and others, that an open market of DNA tests is the twenty-first century's ticket to a healthier nation. But a closer look suggests that this fast-growing industry, with its snazzy, Web-based come-ons, could benefit from some temperance and independent oversight.

The technology is undeniably impressive. For as little as $399, anybody who can drool into a mailing tube can now find out his or her genetic odds of getting any of twenty or more potentially debilitating diseases, including cancer, heart disease, and diabetes. Most of these tests will not lead to a frank diagnosis, as happened with Gulcher. But discovering an inherited propensity toward a particular illness can motivate consumers—or, as they used to be known, patients—to get more frequent checkups, take preventive medicines, or make lifestyle changes to try to ward off the specter of disease. At last, we seem to be on the cusp of the long-promised personalized-medicine revolution in which gene tests allow physicians to craft far more individualized and effective ways of keeping us well.

But tests that look into the fog of people's medical futures are freighted with tricky medical, economic, and bioethical implications. For one thing, most genes are not determinative, so these tests can convey only odds, not destinies. Even with the doubled lifetime risk for cancer that's associated with Gulcher's prostate-gene variant, two out of three men who receive a positive test for that gene will never get the disease. And many of those who do will get it so late in life and in such a benign form that no treatment would be justified. So that's at least two new members of the "worried well" who could be losing sleep and spending money on unnecessary follow-up tests for every person who would arguably be appropriately forewarned.

Moreover, the tests are still new and easily misinterpreted, even by professionals. Online results may be subject to security and privacy breaches. And some companies are using people's gene profiles to conduct independent research. That suggests to many ethicists and lawyers that these firms' paying clients ought to be informed that they are subjects in experiments, with full disclosure of potential risks and rights.

Most worrisome of all, at least a few companies seem to be peddling DNA-based versions of snake oil. Some firms claim to be able to identify inherited nutritional deficiencies that—guess what?—are treatable with pricey supplements that they just happen to sell. Some even promise to discern from your genes what kind of person you should marry to ensure a blissful sex life and healthier babies. Welcome to the Wild West of personalized genomics.

These problems are not insurmountable. But there is precious little oversight of this burgeoning new industry, in part because genetic analysis does not fit cleanly into any existing category of medical practice. And if the first wave of DNA-screening companies to hit the market gets its way, there won't be any more adult supervision in the foreseeable future. In an apparent effort to stave off regulation, top officials from all the major competing gene-test companies convened a series of closed-door meetings in 2008 to hammer out a "best practices" document that spells out the rules of engagement they will follow as they expand their markets. Those rules offer some assurance that the nation's first generation of gene-test customers will not be completely hoodwinked, but they do not go far enough.

One persistent problem is that no state or federal agency can today assure consumers that the DNA tests they order will give accurate results, or that the results, even if technically accurate, will have any practical value. The Food and Drug Administration says it has the authority to regulate all gene tests but has decided to ignore the vast majority of those developed so far. The Federal Trade Commission (FTC), which is supposed to protect consumers from fraudulent claims, has yet to take any enforcement action against even the most transparently deceptive gene-test companies. And the Centers for Medicare and Medicaid Services, the division within the Department of Health and Human Services (HHS) that oversees clinical laboratories, has so far opted to steer clear of the genetic-testing world, despite pleas from federal advisers to ensure a minimal standard of gene-test proficiency.

The companies say that what they do is different enough that they should not be shoehorned into the conventional medical-testing rules. "For the first year and a half of our existence, all we did was try to figure out how to fit into the regulatory environment," said Dietrich

Stephan, co-founder of Navigenics, a leading California-based gene-test company, adding that the effort cost an estimated $10 million.

That's real money. Yet even with all that preparation, Navigenics and a dozen other testing companies recently received warnings from individual states accusing them of violating state rules for labs. Situations such as this cry out for the guiding hand of the federal government—not necessarily through cumbersome regulations, which can be too rigid to keep up with quickly changing science, but through formal guidelines, at least, promulgated by HHS. These could set clear expectations about how accurate gene tests should be—and what it means to be accurate in the brave new world of predictive health—and what level of informed consent should be obtained from clients. The promulgation of such standards will take real effort from the new secretary of HHS. The FTC also needs to show that it has teeth and can bite.

Genetic-testing companies need to ante up, too. The responsible ones could buy a lot of goodwill by offering the public easily accessible scientific details (online and elsewhere) about the specific genes or genetic markers they are testing for; citations for the published studies they use to justify their claims that those genes have real medical relevance; the privacy and security systems they have in place; and the protocols for any experiments that clients' specimens may be used in. The firms should also disclose any approvals they have sought or obtained from independent scientific and ethics review boards, and any applications that have been denied.

At an HHS meeting in mid-2008, the public bore witness to heartening evidence that such a transparent, consumer-friendly future is possible: chiefs from the five major competing gene-test companies sat next to one another, spoke cooperatively, and listened to concerns being expressed by federal advisers. If those executives move aggressively to do the right thing, and if federal officials help them with some smart but tough guidance, perhaps those corporate heads can avoid a future in which they are called upon to appear side by side again—this time before Congress, looking more like those famously photographed tobacco CEOs, being asked tough questions about what exactly they have been selling, and at what cost to American health.

SECURING SOCIETY

National security has been of utmost importance to Americans since the earliest days of the union, but never more so than in the post 9/11 era, a time when the joys of social gatherings and the simplicities of public space have become weirdly overlaid with the presumption of surveillance and an only slightly subconscious awareness of the potential for terror.

For its part, science has made security an ever more challenging goal. Researchers today can stitch together off-the-shelf gene and protein products to make disease-causing viruses from scratch. Suitcase-sized nuclear devices are reputed to be circulating in shadowy corners of the Earth. And, as we have become an increasingly wired society, we also have become aware more than ever of our dependence on those wires, and of the pandemonium that could be wrought by their disruption.

Yet trying to place a cold halt on expanding scientific knowledge and progress is hardly the solution, if for no other reason than it is not under consideration by those who might want to use that knowledge against us. Rather, a progressive view calls for smart research prioritization and good governance over dual-use technologies such as biotechnology and nuclear science. It also requires a recognition that goals not instantly recognizable as having national-security import—such as keeping our transportation infrastructure intact, or fulfilling at last the promise to get high-capacity broadband connections to every city in America—can have revolutionary impacts on our day-to-day safety, in the case of broadband by strengthening the distributive web of communication that would be so crucial in the hours and days after a national emergency.

With proper guidance and intelligent oversight, science and technology can help give future generations the greatest gift of all: freedom from fear.

Microbes and the Militants: Preparing for the Battle of Biotech

I N A CAUTIONARY MESSAGE TO THE WORLD JUST BEFORE HIS RETIREMENT, then U.N. Secretary-General Kofi Annan said that as biological research expands and advanced biotechnologies become increasingly available, the associated safety and security risks will increase exponentially. "When used negligently, or misused deliberately, biotechnology could inflict the most profound human suffering—ranging from the accidental release of disease agents into the environment to intentional disease outbreaks caused by state or non-state actors," he warned. "Soon, tens of thousands of laboratories worldwide will be operating, in a multi-billion-dollar industry. Even novices working in small laboratories will be able to carry out gene manipulation."

Annan then pointed out that the world lacks a system of effective safeguards for managing the risks of biotechnology. "Scientists may do their best to follow rules for the responsible conduct of research," he said. "But efforts to harmonize these rules on a global level are outpaced by the galloping advance of science itself, and by changes in the way it is practiced." To address this policy challenge, Annan called for the creation of a "global forum" under U.N. auspices so that representatives from governments, industry, science, public health, law enforcement, and the public could discuss how to ensure that biotechnology serves the common good.

After Annan's retirement, his successor, U.N. Secretary-General

Ban Ki-moon, decided to take up the proposal and make it a reality. The reason: preventing the misuse of biotechnology for the creation of weapons of almost unimaginable horror will require coordinated global action that only an international forum can muster.

What are the biotechnology-related security concerns that prompted Annan's proposal? One is that many biotech facilities, and the equipment and materials they house, are inherently dual-use, meaning they can be applied either for legitimate civilian purposes or for the development and production of biological weapons. Moreover, whereas biotechnology was once the exclusive domain of advanced industrial countries such as the United States, Germany, and Japan, it is now a major focus of investment by developing countries such as China, Cuba, India, Indonesia, Malaysia, Singapore, South Korea, South Africa, and Taiwan.

There are many good reasons for the spread of biotechnology: it can enhance public health, improve agricultural yields, and foster economic development. Yet the proliferation of dual-use biotechnologies to unstable regions of the world, where war, trafficking of weapons, and terrorism are rife, is potentially a recipe for disaster. In addition to the risk that biotech facilities, equipment, and materials might be diverted to bioweapons production, state or non-state actors could conceivably exploit for hostile purposes certain types of scientific information generated by biomedical research.

Historically, scientists have viewed the discovery of new knowledge as an unalloyed good that contributes to human understanding of the natural world and leads to beneficial applications. But a number of observers, among them bioethicist Arthur Caplan of the University of Pennsylvania, have made the case that with the recent advent of modular technologies and simplified "kits" for altering genes and cells, seemingly benign scientific information can today be quite dangerous in the wrong hands. "We have to get away from the ethos that knowledge is good, knowledge should be publicly available, that information will liberate us," Caplan says. "Information will kill us in the technoterrorist age."

Given these concerns, a philosophical question facing the life-sciences community is whether certain areas of research constitute "forbidden knowledge" that should be banned or otherwise restricted on security grounds. Controversy over the risks of dual-use life-sciences research first erupted in early 2001, when Australian scientists published a paper in the *Journal of Virology* reporting the finding that insertion of a single gene for an immune-system protein into the mousepox virus made this normally benign virus extremely lethal in mice, even those that had been vaccinated against it. Because bioweapons developers could possibly use the same manipulation to increase the lethality and vaccine resistance of related viruses that infect humans, such as monkeypox, critics argued that the information was dangerous and should not have been published.

Weighing in on the debate, the National Research Council (the policy-analysis arm of the U.S. National Academies) convened an expert committee chaired by Gerald Fink, a biology professor at the Massachusetts Institute of Technology (MIT). In late 2003, the Fink Committee released its report, *Biotechnology in an Age of Terrorism*. It concluded that certain types of basic research in the life sciences, although conducted for legitimate purposes, could indeed generate findings that might be misused by others to threaten public health or national security.

The Fink Committee identified seven "experiments of concern" that would render a pathogen more deadly or transmissible; able to infect additional species; resistant to existing vaccines or therapeutic drugs; easier to convert into a weapon; or capable of evading diagnostic or detection techniques. In response to one of the committee's recommendations, the Bush administration established the National Science Advisory Board for Biosecurity, which met for the first time in mid-2005.

The NSABB's mandate is to develop criteria for identifying dual-use research, draft guidelines for the review and oversight of risky experiments, and recommend possible restrictions on the publication of sensitive data. The board consists of up to twenty-five voting members from the U.S. scientific and national-security communities, along with nonvoting representatives from the fifteen federal agencies that conduct or support research in the life sciences.

To give a recent example of dual-use research, in June 2007 researchers at Germany's Helmholtz Center for Infection Research reported in the journal *Cell* that they had altered the DNA of the *Listeria* bacterium, a human pathogen, to enable it to cause disease in mice, a species it does not naturally infect. This finding opened the way to developing a mouse model of *Listeria* infection, a key step in developing new treatments for the disease.

Yet the experiment has troubling security implications because the technique used to modify the host range of the bacterium—that is, the range of species it can infect—could potentially be applied in reverse, enabling an animal pathogen to infect humans. Despite this risk, the editors of *Cell* did not seek outside advice about whether to publish the study. Moreover, because the German researchers could have published their work in a European journal, this case suggests that U.S. controls on dual-use research will not be effective unless other countries sign on.

Another emerging area of biotechnology with security implications is synthetic genomics, which involves the design and synthesis of long strands of DNA. The DNA molecule encodes genetic information with an alphabet of four "letters," or nucleotide bases (abbreviated as A, T, G, and C), which can be strung together in any conceivable sequence. The advent of automated DNA synthesizers has spawned a new industry in which hundreds of companies around the world—including firms in China, India, and Iran—synthesize DNA sequences to order. A researcher seeking a particular piece of DNA simply goes to the supplier's Web site and enters the desired nucleotide sequence and a credit card number; several days later a vial containing the synthetic DNA arrives in the mail.

Most of the market is for small snippets of DNA for basic research. But a small fraction of DNA-synthesis companies, known as gene foundries, are capable of making gene-length strands of DNA consisting of thousands of nucleotide base pairs. These segments can then be assembled in the right order to form an entire genome—the gentic blueprint of a microorganism. Since 2002, scientists have used this technique to reconstruct two human viruses in the laboratory: poliomyelitis virus (7,440 base pairs) and the formerly extinct Spanish influenza virus (13,500 base pairs), the latter of which killed tens of millions of people during the worldwide pandemic of 1918–19. Both synthetic viruses have been shown to be infectious and capable of causing illness in experimental animals. Given the rapid advances in automated DNA synthesis, it is only a matter of time before it becomes possible to synthesize larger viruses in the laboratory, such as Ebola virus (about 19,000 base pairs) or even smallpox virus (about 186,000 base pairs). This development would make it possible to circumvent the physical security measures that currently keep such deadly pathogens out of the wrong hands.

A more ambitious goal of synthetic biology is to create novel genetic circuits that would enable microorganisms to perform practical tasks, with applications in medicine, computation, environmental remediation, and energy production. Futuristic examples include giving bacteria the ability to sequester carbon dioxide or to manufacture hydrogen fuel. To facilitate this task, Professor Drew Endy, now at Stanford University, worked with colleagues at MIT's Department of Biological Engineering to compile a tool kit of pieces of DNA with well-characterized functions. These components can be assembled into functional genetic circuits, much as electronic devices are built from transistors, resistors, and diodes.

To date, synthetic biologists have demonstrated the basic concept by performing a series of ingenious parlor tricks. For example, they have designed genetic modules that cause bacteria to blink on and off like microscopic Christmas-tree lights, or to become light-sensitive so that a lawn of the bacteria behaves like a photographic plate. Similarly engineered life forms could someday help solve some of humankind's biggest environmental and energy crises.

Despite the potential benefits of synthetic biology, however, the field could provide individuals with malicious intent with new ways to cause harm. For example, it may become possible to engineer novel viral or bacterial genomes capable of expressing toxins or virulence factors for which no natural immune defenses exist, and against which existing therapeutic drugs are powerless. In addition, as synthetic biology diffuses widely, a new breed of "biohackers" might emerge, intent on showing off their prowess by developing real viruses rather than digital ones.

The potential misuse of biotechnology for hostile purposes is not limited to the development of more deadly microbial and toxin agents. Following on the work of the Fink Committee, another National Research Council panel, co-chaired by virologists Stanley Lemon and David Relman, issued a report in early 2006 titled *Globalization, Biosecurity, and the Future of the Life Sciences*, which concluded that several other areas of biotechnology and biomedical research also pose dual-use dilemmas.

One case in point: Advances in drug-delivery systems, such as needle-free systems for administering insulin to diabetics in the form of an inhalable aerosol, have security implications because aerosolization is the optimal method for disseminating biowarfare agents over large areas.

What can be done to manage the risks of dual-use research in the life sciences without causing significant harm to the scientific enterprise, which thrives on the open sharing of information? The U.S. National Science Advisory Board for Biosecurity has defined "dual-use research of concern" as:

> Research that, based on current understanding, can be reasonably anticipated to provide knowledge, products, or technologies that could be directly misapplied by others to pose a threat to public health and safety, agricultural crops and other plants, animals, the environment, or materiel.

This definition sets the threshold fairly high: The risk of misuse must arise directly from the research findings and have significant implications for public health, agriculture, or national security. For example, an experiment that creates a highly virulent organism, but one that cannot be readily transmitted to humans, would not be considered a major threat.

The NSABB has developed draft guidelines for the oversight of federally funded research in the life sciences to minimize the risk of misuse for hostile purposes, but national biosecurity measures are not sufficient. Because biotechnology is a global activity, managing its downside risks will require adopting policies at the international level to ensure that only legitimate scientists have access to deadly pathogens and to oversee potentially dangerous research. At present, biosecurity rules vary widely from country to country, creating a regulatory patchwork with gaps and vulnerabilities that bioterrorists could exploit as targets of opportunity. Moreover, if other countries adopt weaker guidelines than those of the United States, then the anticipated security benefits of the U.S. regulations will not materialize, and American researchers and scientific journals will find themselves at a competitive disadvantage.

Any effective global system for regulating biotechnology will have to be based on common standards for laboratory security and research oversight, which countries would implement and enforce on a national basis. In 2006, the World Health Organization took a useful first step in this direction by issuing a set of guidelines for securing dangerous pathogens in locked cabinets, vetting laboratory personnel to make sure they are bona fide scientists, and keeping accurate records, but only some research labs have adopted these rules.

More action is clearly needed. To prevent the misuse of biotechnology, the Lemon-Relman report called for creating a "web of prevention" encompassing individual scientists, their research institutions, national governments, and international agreements. Many of the key elements of this web already exist as a result of ongoing efforts to prevent biological warfare and can be found in the Biological and Toxin Weapons Convention of 1972, transnational networks of scientists and other stakeholders, export-control regimes, professional codes of conduct, and educational and awareness efforts.

It is unclear, however, what institutional mechanism could serve to coordinate biosecurity measures at the global level. Although the Vienna-based International Atomic Energy Agency inspects civilian nuclear plants around the world to ensure that fissile materials are not diverted for nuclear weapons, that approach is a poor fit with biotechnology. The nuclear industry consists of a limited number of facilities and stocks of radioactive materials that are amenable to precise accounting, yet the biotechnology industry is extremely diffuse and involves the use of self-replicating organisms that cannot be tracked in a quantitative manner.

Harvard University biochemist Matthew Meselson warns that twenty-first-century biotechnology will make it possible not only to devise additional ways to destroy life, but also to manipulate the processes of cognition, development, reproduction, and inheritance, creating "unprecedented opportunities for violence, coercion, repression, or subjugation." He says that "movement towards such a world would distort the accelerating revolution in biotechnology in ways that would vitiate its vast potential for beneficial application and could have inimical consequences for the course of civilization."

Given these very real dangers and the complex challenges of developing an international mechanism to manage the risks of dual-use research in the life sciences, the UN Secretary-General's global forum on biotechnology will provide an important vehicle for addressing one of the major security challenges of our time.

Bringing Broadband Up to Speed

THE UNITED STATES DID NOT MEET PRESIDENT BUSH'S GOAL OF UNIVERSAL broadband by the end of 2008—not by a long shot. The number of subscribers to Internet services grew substantially as the administration came to a close, and continues to grow today. Yet for the most part these subscribers are not connected to broadband technology, which Congress described in 1996 as a two-way communications service capable of high-speed delivery of data, voice, and video.

This failure to connect over half the country to advanced telecommunications service is not a technological failure. It is a twenty-first-century public-policy failure. In the 1990s, policies established by the Clinton administration to encourage public/private telecommunications partnerships, to connect schools and libraries to the World Wide Web, and to allow competitive service providers onto the networks of the local telephone monopolies all sped up the deployment of broadband around most of the nation. These policies were either deliberately abandoned or hampered by the Bush administration.

The increasing noise from Washington about the lack of a U.S. broadband policy obscures the fact that a policy choice was made by the Bush administration to rely entirely on market forces to determine how and where advanced telecommunications services would be deployed. That policy failed, and it is time to reassess the situation so progress can begin.

The goal of federal investment in broadband should be first and foremost to ensure our ability to respond to threats to our homeland security and to natural disasters. And the result of administration neglect, industry intransigence, and the incompetence of a Federal

Communications Commission apparently captured by the industry it is supposed to regulate has left the American people and most policy-makers with no clear idea where broadband services are deployed in the United States.

There is no credible dispute that the United States has fallen behind Canada, France, Japan, and a dozen other industrial countries in broadband deployment. Americans are not more averse to new technology compared to our neighbors to the north or our friends overseas. The difference is that these countries have moved ahead of the United States after having adopted one version or another of U.S. telecommunications policies established in the mid-1990s.

In addition to leaving America less competitive in a global economy, this failure has left the nation vulnerable and ill-prepared for real threats to our national security—the rationale behind the initial U.S. government investment in the development of the Internet.

The American invention of the Internet, of course, was preceded by hefty scientific investments for military purposes, beginning with the Eisenhower administration. In fact, the Internet developed despite market forces dominated by the not-so-invisible hand of the Bell telephone monopoly. While the development of the Internet has certainly benefited from global market forces, the free-market blinders that prevent present-day U.S. policymakers from putting the national interest ahead of corporate interests must be removed. Moreover, this is not a partisan issue—although Reagan-era Republicans do seem to don those blinders with greater pride. It was, after all, Vice President Al Gore who insisted that the information superhighway would not be built the way the U.S. highway system was built, but would instead be financed by private enterprise.

If the United States is to catch up with other developed and developing nations, however, we must look beyond even the abandoned policies of the Clinton era and begin to move with greater urgency and resolve to address pressing disaster-response and defense needs. After all, the attacks of 9/11 and the body blow of Hurricane Katrina highlight for all but the most doctrinaire advocates of free markets that there is an exceedingly strong case for direct government investment in the deployment of advanced telecommunications services to build a safe, strong, and resilient America.

While the goal of federal investment in broadband should be first and foremost to ensure our ability to respond to threats to our homeland security and to natural disasters, there is a second, closely related goal: the availability of advanced telecommunications services in our health-care and educational systems. The modernization of these systems will be key to our nation's ability to respond to threats to our national security and public safety, both immediately and over the coming decades. Without ubiquitous broadband our first responders could be crippled by the lack of effective communications in the event of a terrorist attack or natural disaster. Similarly, our educational institutions need to be able to communicate quickly and effectively in case of a pandemic, as well as conduct R&D on all of the technologies needed to maintain our nation's national defense and public safety.

In meeting these goals, federal investment should make certain that the U.S. communications infrastructure is continually upgraded, robust, redundant, and able to withstand multiple threats and uses. The public should not be left to rely on any one technology, but rather should have access to multiple technologies—each able to operate with the other, and each able to serve important needs if the other technologies are destroyed or compromised. Market forces will not guarantee this result.

In small rural towns, in the crowded barrios and ghettos of urban U.S. cities, in those places where financial institutions are not yet convinced they can get an adequate return on investment, Americans do not have access to the communications networks they will need to keep them safe in the future. It is no coincidence that these same places hold many of our nation's toxic waste dumps, our chemical plants, and our seaports and airports—places whose destruction by terrorism or natural forces would cause immense disruption. Yet we do not have the ability to communicate most effectively in these most vulnerable places.

The Department of Defense has long been provided almost all the communications resources it needs to protect American interests overseas. What has been too often forgotten is the importance of equipping all Americans with the ability to participate effectively in the national defense effort at home. Americans take pride in assisting when their communities are under attack or threatened by a natural disaster. A concerted effort must be made to equip all Americans so they are able to communicate effectively when confronted by catastrophe.

President Dwight Eisenhower understood the value of a robust transportation system at home to sustain national unity and to promote defense needs. In announcing the new Interstate Highway System, Eisenhower called the effort "the National Defense Highway System," citing his direct experience with a problem-laden military convoy from Washington, D.C., to San Francisco he took in 1919.

Despite the squabbles of some local government and business leaders who fought against a federal highway system, Eisenhower was convinced that America could do better. As Richard Weingoff reports in his excellent history of the interstate system, when Vice President Richard M. Nixon delivered an address before a 1954 conference of state governors at Lake George, N.Y., reading from Eisenhower's detailed notes, he declared that the U.S. "highway network is inadequate locally, and obsolete as a national system."

Nixon then recounted Eisenhower's convoy and cited five "penalties" of the nation's obsolete highway network: the annual death and injury toll, the waste of billions of dollars in detours and traffic jams, the clogging of the nation's courts with highway-related suits, the inefficiency in the transportation of goods, and "the appalling inadequacies to meet the demands of catastrophe or defense, should an atomic war come."

If America is to be ready "to meet the demands of catastrophe or defense," all Americans need access to advanced telecommunications services in the twenty-first century, just as they needed access to an advanced highway system in the twentieth century. But as the 9/11 Commission noted in its report, the United States is not ready for a national emergency. And as every comprehensive analysis of the tragedy of Hurricane Katrina revealed, we are not prepared to handle a major natural disaster. Both of these experiences highlight the importance and the multiple failures of U.S. communications services as warning systems or as systems to allow for the coordination of first responders.

A standard complaint of conservative defenders of the current telecommunications regulatory system regarding communications policy focuses on the supposed "command and control regulatory policies" of the federal government. They argue that the heavy hand of regulation stymies the rollout of advanced telecommunications networks across the nation, when in fact the tendency of the federal government historically is to exercise this command and control on behalf of the communications industry itself.

The result of this regulatory protection of different bits of the telecommunications industry leaves the United States with balkanized communications capabilities. If the prevention or response to the terrorist attacks on 9/11—when New York City police, fire, and rescue workers could not communicate with each other amid the chaos and carnage of that awful day—or the prevention or response to the failed levees overwhelmed by Hurricane Katrina demonstrated anything, it was the need for more, not less, command and control.

Indeed, in the debate over communications policy, the term "command and control" is little more than a right-wing slogan. Outside of military operations this phrase has never accurately described either the policymaking process or the execution of policy in the United States. Even the federal highway system, so important to Presidents Franklin Roosevelt, Harry Truman, and Eisenhower for military purposes, was the product of a contentious federal-state partnership.

Still, there is no question about the importance of federal vision, leadership, and funding. The importance of strong federal engagement in the development of the national highway system is beyond dispute. The same can be said of the importance of federal leadership in the U.S. space program, which led to the U.S. satellite industry, as well as federal leadership in the Defense Advanced Research Projects Agency, which spurred the research behind the Internet.

Perhaps the most direct corollary to the national highway system in the U.S. telecommunications arena is the National Communications System. The NCS began after the Cuban missile crisis. Communications problems between and among the United States, the Soviet Union, and other nations helped to create the crisis. President John Kennedy ordered an investigation of national-security communications, and the National Security Council recommended forming a single, unified communications system to connect and extend the communications network serving federal agencies, with a focus on interconnectivity and survivability.

The NCS oversees wireline (Government Emergency Telecommunications Service) and cellular service (Wireless Priority Service). The NCS is now part of the Department of Homeland Security's Preparedness Directorate, and despite the increased attention to the communications needs of first responders on September 11, 2001, NCS failures and inadequacies were made obvious after Katrina. In New Orleans, police officers were forced to use a single frequency on their patrol radios, which "posed some problems with people talking over each other," explained Deputy Policy Chief Warren Riley at the time. "We probably have 20 agencies on one channel right now." And with little power to recharge batteries, some of those radios were soon useless.

In southern Mississippi, the National Guard couldn't even count on radios. "We've got runners running from commander to commander," said Maj. Gen. Harold Cross of the Mississippi National Guard. "In other words, we're going to the sound of gunfire, as we used to say during the Revolutionary War." As Senator John Kerry (D-MA) said, "This is a further demonstration of our inadequate response to the 9/11 Commission's recommendations and other warnings about the failures in our first responders' communications systems."

How can these obvious communications failures still leave the United States groping for an adequate response? One of the biggest challenges we face is the tendency to see national defense and emergency needs regarding communications as separate and unrelated to the communications needs of the American public. The NCS has established an elaborate set of protocols that make government communications a priority over what is called the public switched network. Federal, state, and local governments pay substantial fees to use this communications network. But how that network is upgraded and deployed is entirely determined by private industry.

We cannot have a robust, survivable, interoperable communications system that protects the public if the public is treated merely as a mass of consumers and not as an integral part of national defense and emergency response. The U.S. public remains vulnerable because our communications infrastructure is too often viewed only as a private business. Katrina and 9/11 remind us that access to advanced telecommunications service is a public need. We need national leadership to remind us of this and insist on policies that address public needs.

In the 1996 Telecommunications Act, Congress indicated that advanced information and communications technology, or ICT, should provide the ability to send and receive data, voice, and video. Today, advanced ICT means the ability to send and receive high-definition video in real time, something that requires massive telecommunications power if the goal is for everyone to be able to do so. Further complicating this goal is that in emergency situations, communications systems become easily overloaded as people rush to their phones to check on loved ones.

In the case of an emergency or national disaster we need a capacity far greater than the market would support for even heavy shopping days. A starting point would be symmetrical speeds (both download and upload capability) of ten gigabytes per second. Today, speeds of that magnitude are available only at the most important point-to-point interchanges of the Internet backbone or between dedicated military, financial, educational, or scientific institutions. Both fiber and robust wireless services have the potential to deliver these speeds in both directions.

But the construction of one or even two robust communications pipelines into police stations or military posts would still leave the United States vulnerable. The sole reliance on only one or two sources of communications creates an inviting target and, at the very least, creates the potential for deadly communications bottlenecks. Telecommunications businesses won't help us solve this problem. At their best, they work to create greater efficiency by *eliminating* redundancy. At their worst, they work to eliminate any and all competition so that even efficiency doesn't matter.

When reliability is essential, redundancy is highly valued. When lives are at stake, establishing alternative systems that can do as good a job as any designated primary system is routine. And while our policymakers speak of competition—sometimes even embracing competitive communications infrastructures that might lead to alternative "consumer" choice—policymakers rarely seem to understand that alternatives are essential to national defense and emergency preparedness.

In fact, redundancy is so essential to public safety and national security that where private industry refuses to create these alternatives government must do so. Engineers consider redundancy a critical ingredient of creating a system with a high probability of safety. In the commercial aircraft industry, for example, pilots and passengers are assured of safety in part because redundant equipment, including engines and sensors, are required by government regulation.

In addition to redundancy, it is vital that the different systems and the equipment operating over these communications systems be interoperable. One unfortunate result of relying on private competition is the tendency of competitors to develop systems that do not permit interoperability. A key failing of emergency response after 9/11 and Katrina was the lack of interoperable communications equipment.

Many of the problems of interoperability are the result of turf wars and not equipment limitations. Federal policies to override local turf wars are essential. The Department of Homeland Security has made it a priority to solve the range of problems related to interoperability. But again, interoperability must not be limited to operation over one infrastructure, but must cross all relevant communications platforms. Phones and computers must operate over wireline and wireless infrastructures, including competing wireline and wireless networks. Interoperability is a vital component of emergency service and a modern communications network. Closed "private" broadband networks stifle not only innovation and service competition, they also limit the ability of all Americans to participate effectively in response to natural disaster and terrorist attack. If the United States is to compete effectively in a global economy and defend itself against global terrorist threats, then it must take advantage of the unique opportunities only possible with an open network.

Federal law should require that all broadband networks be open to the attachment of any equipment the user chooses—so long as it does not harm the technical operation of the broadband network. In addition, federal law should require broadband networks to be open to other information-service providers and accessible to other networks, except for restrictions related to vital law enforcement or for network management.

Our nation's wireline infrastructure is inadequate to meet twenty-first-century needs. The old telephone network is simply incapable of delivering the bandwidth to meet the emergency needs of today and the future. While efforts have been made to upgrade the relatively more modern cable infrastructure, there are too many rural communities where the cable system has not been upgraded to provide digital service. Even in our major metropolitan areas, gross deficiencies are self-evident.

The strain on the existing telecommunications infrastructure was obvious as call after call was blocked during 9/11. But this strain is also obvious to anyone who regularly uses either the Internet or regular cell-phone service in a major metropolitan area in the United States. The concerns that the Internet as presently constructed simply will not bear the amount of use projected over the next five years are long-standing. While more sophisticated filtering and better emergency protocols may address this problem in the short term, the strain on the nation's telecommunications infrastructure will only increase as the call for greater bandwidth for video over the Internet increases.

If meeting the communications needs of first responders or panicked parents were simply a matter of market forces, then one would be tempted to applaud the telephone and cable companies for squeezing as much profit as possible out of old technologies. But the challenge of communicating in an emergency should not be held hostage to even legitimate profit-seeking demands of private investors.

In brief, the nation should be investing in the deployment of opical fiber, power-line, wireless, and satellite communications technologies. The combination of these technologies would ensure robust and ready communications services in case of a national emergency. What's more, these technologies are readily available for rollout, as we will detail below.

Optical Fiber

The most promising single technology that could deliver advanced telecommunications connectivity to homes and offices everywhere is optical fiber, a thin glass or plastic line designed to distribute light. Optical fiber is distinct from the electricity that distributes communications through copper telephone wires or coaxial cable. The light in optical fiber permits transmission of digital data over longer distances and at higher rates than other forms of communications.

Fiber-optic products have been used for several decades in a variety of defense technologies designed for air, sea, ground, and space applications. During the high-technology boom of the 1990s many privately held companies and public corporations built out vast fiber-optic networks even as telecommunications companies, beginning in the early 1990s, started to upgrade their networks to incorporate fiber technology. Yet only one large U.S. company, Verizon, has extended optical fiber to the home.

The immediate reaction from Wall Street to Verizon's plans was pessimistic. Verizon's stock value dropped in 2006, and investors pressured the company to scale back deployment or abandon the investment in fiber to the home altogether. Investors saw little reason to back Verizon's expensive ($23 billion) proposition.

Never mind that over time Verizon's emphasis on delivering video entertainment alongside other telecommunications services so the company could compete with cable is now increasingly viewed as smart, forward-thinking investment strategy. Unfortunately, Verizon's service areas are largely densely populated urban areas, and Verizon's rural customers are not likely to get fiber anytime soon. And while a few small regional operations are also deploying fiber to the home, Verizon remains the only major telecommunications company doing so. Again, the emphasis on market priorities, forward thinking or not, does not serve the goal of protecting Americans with the best communications service available in case of an emergency.

A growing number of municipalities and public utilities, however, have deployed or are planning to deploy fiber networks. Some are investing in fiber with the expressed intent of improving the communications capability of emergency workers. One example is Arlington County, Virginia, just across the Potomac River from Washington, D.C. Arlington firefighters were the first to respond on September 11, 2001, when the Pentagon was attacked by terrorists. Beginning with its ten fire stations in January 2002, by June 2002 all forty county sites were connected to a fiber network. In 2005, Arlington extended the network to the nearby city of Alexandria, to facilitate interagency collaboration.

These are the kinds of public investments that federal, state, and local governments all need to make in tandem with the private sector to ensure that households and offices are all connected to the most readily available form of high-speed telecommunications. Ubiquitous broadband via fiber optics is the best first step that could be made by such a public/private partnership.

Power-Line Communication

Broadband over power lines, known as BPL by industry insiders, is a second promising technology in need of expanded deployment. It would make use of the extensive electrical power-grid infrastructure to communicate digital signals. Unlike fiber-optic systems, however, BPL still has some kinks to be worked out, because both the electric grid and the home exist in what engineers call a noisy environment. For example, every time a device turns on or off, a pop or click is introduced into the line.

BPL has developed faster in Europe than in the United States because of differences in power-system design philosophies. Large power grids transmit power at high voltages to reduce transmission losses, and transformers that are near the customer reduce the voltage. Because BPL signals cannot pass through transformers, repeaters must be attached to each transformer. In the United States, a small transformer typically services a single house or a small number of houses. In Europe, it is more common for a larger transformer to service up to 100 houses. Delivering BPL over the power grid of a typical U.S. city will require many more repeaters compared to a typical European city.

Despite these challenges, BPL in the United States is on the rise, with about 6,000 BPL subscribers nationwide as of 2006. According to the United Power Line Council, commercial deployments are up slightly, from six in 2005 to nine in 2007. Trial rates, however, have fallen from thirty-five in 2005 to twenty-five in 2007.

An indication of a possible increase in BPL penetration, however, came in 2007 when DirecTV announced it was getting in on the BPL market. In a deal with Current Group, DirecTV plans to provide BPL service in the Dallas-Fort Worth and Cincinnati areas with a potential for much broader rollout. Not to be outdone, Oncor, a subsidiary of Dallas power company Energy Future Holdings Corporation—formerly TXU Corporation—has started to deliver BPL service, and it recently passed 108,000 customer deployments, less than 5 percent of its goal.

The rise in BPL deployment can also be traced to steps the FCC took in 2006 to support the technology by reaffirming an earlier decision that BPL providers have the right to provide data access using power transmission lines so long as they do not interfere with existing radio service. Still, opponents of BPL, including the aviation industry and the amateur-radio community, have continued to voice the strongest concerns over the issue of possible interference with radio communication, though there is some dispute among experts over the degree to which electricity over BPL actually "leaks" and thus interferes with an electromagnetic wireless signal.

In a further boost, the FCC classified BPL-enabled Internet access as an information service, rather than a telecommunications service, in November 2006. According to the FCC, "The order places BPL-enabled Internet access service on an equal regulatory footing with other broadband services, such as cable modem service and DSL Internet access service." According to Joe Marsilii, president and CEO of BPL equipment-maker and integrator MainNet Powerline Inc., 70 to 80 percent of the nation's electrical grid will be equipped with BPL in five to eight years.

This kind of rollout of BPL services, however, will not occur without a coherent policy advanced by those federal agencies responsible for keeping America competitive and secure. BPL could easily become the second ubiquitous source of broadband to all houses and offices with a plug. With only a few technology hurdles to clear, and with FCC regulatory clearance already evident, BPL through a public/private partnership could become available swiftly.

Wireless Broadband

Wireless offers a third potential level of coverage, but as anyone who has attempted to carry on cell phone conversations in New York City or rural America will attest, reliance on the most prevalent wireless technology in America would be misplaced. Cell phones are no less ubiquitous in big American cities than they are in London or Taipei or Toronto, but somehow cell phones seem much more reliable in other countries.

Coverage problems in the United States result from the lack of cell-phone infrastructure—towers and repeaters—necessary to sustain a large number of users in the variety of locations. The infrastructure problems are directly tied to two factors. First, the costs to build that infrastructure at present outweigh the commercial benefit, which is the profit the telecommunications companies and their shareholders think they can realize. Second, because cell-phone service is seen only as a commercial need, there is little public will to assist in supporting the cost of this infrastructure development by allowing, mandating, or helping to finance the build-out of towers and repeaters.

Coverage problems also result from the limited propagation characteristics of the spectrum set aside for cellular service. Most cell phone use in the United States is based on dated technology. Advanced digital Internet protocols make possible voice, data, and video communications over mobile networks. Third-generation, or 3G, broadband has been deployed effectively in the United Kingdom, Germany, Japan, and other countries, but the United States lags.

The creation of a next-generation wireless broadband network is an important public-policy goal. The public-safety benefits of reaching this goal justify significant federal funding to subsidize the development of such a network. One proposal is that the funding of a 3G public-safety network could come by redirecting the billions of dollars designated to the federal government's wireless-network project—estimated between $5 billion and $10 billion—and which will only serve a limited number of federal agencies.

The focus, however, should not be on any one technology, but rather on the full funding of a public-safety network that utilizes wired and wireless infrastructures. The establishment of a public-safety network can serve as a strong starting point for the development of a next-generation network for commercial purposes, but a public-safety network should not be held hostage to commercial interests.

Federal allocation of spectrum must be revised to allow for the deployment of advanced wireless technologies. Licenses for all current analog radio and television broadcasting should be revoked in 2009.

Google, Microsoft, and others recently successfully argued that the FCC should open up to unlicensed applications the "white spaces" in the soon-to-be-abandoned analog broadcast frequencies. White spaces are the vacant television channels that were set aside to prevent television broadcasters from interfering with each other. For example, if a local community had channels four, seven, and nine, channels three, five, six, and eight could not be used in that area or in nearby communities. These white spaces were to become available in VHF channels three through fifty-one (channels two and thirty-seven are reserved for radio astronomy and medical telemetry) beginning in February 2009, when full-power analog TV stations were scheduled to cease operations.

The portion of the television-broadcast band that is white space in each community ranges from 20 to 30 percent in highly congested urban markets to 70 percent or more in small city and rural markets. For example, in San Francisco, there are nineteen vacant channels, leaving 37 percent of the broadcast-television band vacant. In Columbia, S.C., there are thirty-five vacant channels, leaving 70 percent of the television band vacant. The part of the spectrum used for analog television has especially powerful transmission characteristics. VHF propagation characteristics are ideal for relatively short-distance terrestrial communication, are less affected by atmospheric noise and interference from electrical equipment than lower frequencies, and are less bothered by buildings and trees than UHF frequencies.

Standards that will allow unlicensed use of vacant VHF channels have been developed and tested, though there remains some controversy over whether devices that would use the proposed standards will interfere with digital broadcasters and others who use that spectrum space. Mark McHenry, chief executive of Shared Spectrum, vigorously disputes the claims of broadcasters and the wireless microphone industry that a broadband wireless protocol using the vacant white spaces would necessarily interfere with existing applications. His company has developed a technology that combines white-spaces standards with smart radio.

Smart-radio technology scans, locates, and then utilizes unused spectrum. McHenry's company developed technology that will allow the Department of Defense to operate wireless broadband in another country without interfering with the existing broadcast infrastructure, and he argues that the debate over white spaces is limited by political considerations, not technical ones. He thinks the United States. should allow for even more-powerful transmission over unused spectrum space than Google and others are calling for.

With the recent FCC decision to reserve this unused spectrum for new uses, and pending adoption of new protocols and standards, including smart-radio technology, a wide variety of innovative applications can now be promoted, including those that would improve wireless broadband communications in emergencies. The transition to digital television should be used as an opportunity to open up powerful broadcast frequencies for public-safety purposes. Half of the vacant spectrum should be reserved for temporary experimental applications with a priority placed on those applications that serve public safety, health care, or educational institutions.

Wi-Fi and Wi-Max

Wi-Fi is a digital wireless communications technology. The brand is owned by the Wi-Fi Alliance, a consortium of companies that have agreed to a set of interoperable products based on a standard (802.11) set by the Institute of Electrical and Electronics Engineers (IEEE). Though the Wi-Fi Alliance apparently originally intended the name to stand for Wireless Fidelity, later statements from the consortium suggest the name is not an acronym or abbreviation.

Wi-Max is an acronym for Worldwide Interoperability for Microwave Access. This was adopted by the Wi-Max Forum in 2001. Wi-Max adheres to the IEEE 802.16 standard and allows for higher-speed networking across much wider geographic distance than is now possible with Wi-Fi. Both Wi-Fi and Wi-Max in the United States face the technical challenges of limited spectrum allocation, particularly when compared with Europe.

More than 300 counties and municipalities have wireless networks. These networks are used for applications ranging from reading meters to managing traffic and providing Internet access. Most municipalities contract with private companies to build and operate the network, and understandably the private industry is primarily concerned about profit. Therefore, in addition to the technical challenges in the United States, there are substantial difficulties with the business model.

Because of the technical and business challenges, large-scale municipal wireless projects are flopping in big cities all across the United States. The problems arising in Houston, Chicago, St. Louis, Philadelphia, and San Francisco are for the most part very similar: the infrastructure (nodes and towers) was not in place, and when private companies were contracted to build the infrastructure, raising public money was difficult. Plans to migrate to public from private service were complicated by the fact that the slower and less reliable Wi-Fi connections are not able to compete effectively against incumbent wired (cable or DSL) Internet providers.

The success stories of municipal Wi-Fi—more than 110 such systems are in operation—come mostly from small towns. In St. Cloud, Fla., a truly citywide municipal Wi-Fi network exists at no cost to residents. Mountain View, Calif., has a citywide wireless network owned by Google with free service to residents. Both these networks operate over relatively small geographic areas: Mountain View is fourteen square miles; St. Cloud is twelve square miles. Most of the American cities and counties attempting municipal Wi-Fi cannot offer it for free.

The telecommunications industry nonetheless argues that the involvement of municipalities creates unfair competition for private organizations because of their ability to use public assets. The industry also argues that municipal governments do not have the necessary expertise to operate or maintain the technology, and anyway should not be "picking winners" in a competition among technological alternatives.

Preoccupation with these industry concerns largely obscures the needs of public safety and emergency response. While neither Wi-Fi nor Wi-Max will address all the communications needs of local communities, the establishment of these systems can help fill in the deployment gaps and assist in providing the important redundancy demands of emergency communication. Fixed microwave wireless communications systems can also help fill in critical gaps. The real problem is the tendency to look for easy answers rather than implement comprehensive solutions that should include Wi-Fi and Wi-Max. Federal leadership is needed to push forward a rationale for public investment that puts a priority on safety and emergency response.

Satellite Broadband

Satellites in geostationary orbit can relay Internet speeds of about 0.5 megabits per second to the user. But satellite broadband typically allows for only 80 kilobits per second from the user. In many rural areas this is a substantial increase over what is typically available. Although DirecTV and a few others have invested in making satellite broadband service a commercial competitor, it suffers from serious competitive disadvantages. Bad weather and sunspot activity can cause unreliable signals and dropouts. Applications such as virtual private networks and voice-over Internet protocol, or Internet telephony, are discouraged or unsupported. And most satellite Internet providers abide by a fair-access policy, limiting a user's activity, usually to around 200 megabits per day.

Perhaps the greatest commercial disadvantage, however, may be the delay that results from the 44,000 miles a signal would need to travel from the user to the satellite company. This delay results in a connection latency of 500 to 700 millisecondss, as compared with a latency of 150 to 200 ms typical for terrestrial Internet service providers.

Still, new technology has decreased the weight and size of satellite antennae and receivers, which, combined with computer tracking devices, makes it easier to send and locate satellite signals. And perhaps the biggest advantage of satellite broadband, particularly for emergency use, is that it can be established quickly on a mobile unit that can avoid an attack or be rushed to the scene of a natural disaster. Fixed towers and telecommunications conduits necessary for wired or terrestrial wireless services are much more vulnerable to attack or natural disasters.

All these communications technologies—satellite broadband, Wi-Fi and Wi-Max, wireless broadband, power-line communications, and optical-fiber networks—are available for local, state, and national governments to warn and protect citizens. It is not a matter of choosing one or the other, but intelligently investing in all these technologies and engaging in research to develop more. Government protection of the U.S. telecommunications industry should take the form of ensuring that the industry is protected in case of an attack or natural disaster; it should not take the form of protecting industry profit at the expense of national security. America needs a robust communications system for emergencies the nation will surely face in the future.

All government offices, health-care centers, primary and secondary schools, military, police and fire, and emergency responders need access to advanced information and communications technology (ICT) to prepare for and respond effectively to natural disasters and terrorist attacks. Federal and state governments may bicker over their relative access to advanced ICT, but there is little disagreement over the need for access. Similarly, while there are disputes on the edges, there is a consensus that police, fire, and emergency responders need this access.

But other institutions in this country require ubiquitous broadband access in order to help our citizens in times of crisis, the two most critical sectors being educational and health-care institutions.

Health-care Instititutions

Health-care centers face extraordinary burdens during and after emergencies. The victims of natural disasters or other catastrophes require medical attention, as do the emergency responders who risk their lives. The ability to diagnose and monitor patients, to access patient records, and to communicate with pharmacists is increasingly dependent upon reliable communications systems within and beyond the hospital.

The absence of robust and redundant communications systems in our community health-care facilities puts at risk not only patients but those who risk their lives to keep the rest of us from having to enter the hospital. In addition, advanced telecommunications systems have proven to be effective in providing access to medical expertise, even over great distances.

A cardiac patient in a small military hospital in Guam, for example, was able to undergo a life-saving heart operation supervised by an expert doctor 3,500 miles away at Tripler Army Medical Center in Honolulu. The surgery was relatively routine for Dr. Benjamin Berg, who was able to dictate the procedure to a less experienced colleague, monitoring every move and heartbeat with a high-resolution video camera and instant sensor gathering data from the catheter as it was slid carefully into the right chamber of the patient's heart.

"The real-time information requires a continuous broadband connection," Berg said. "The delay in the transmission of data about pressure inside the heart would be unacceptable." Imagine doctors being able to help patients remotely as the health-care centers in New York and the Gulf Coast were inundated.

The example cited above of the surgeon in Honolulu supervising an operation in Guam is but one of the remote-care practices engaged in by the Department of Veterans Affairs system. The VA also works with the Alaska Federal Healthcare Access Network, which links nearly 250 sites, including military installations, Alaska Native health facilities, regional hospitals, small village clinics, and state of Alaska public health nursing stations, to provide various health-care services, using high-speed broadband services, including satellite broadband.

A VA study of a remote-monitoring program demonstrated a 40 percent cut in emergency room visits and a 63 percent reduction in hospital admissions. A separate Pennsylvania State University study estimated that remote home-health monitoring for diabetes patients cut costs for hospital care by 69 percent. According to Jon Linkous of the American Telemedicine Association, "Broadband Internet access to hospitals is becoming a critical tool in the delivery of medical services."

In addition to providing the communications infrastructure to local health-care facilities, it is vital to increase support for the National Institutes of Health (NIH) and the Centers for Disease Control and Prevention (CDC). NIH has long demonstrated its importance in emergency and disaster readiness. One notable program is the University of California, San Diego, and the California Institute for Telecommunications and Information Technology's $4 million WIISARD (Wireless Internet Information System for Medical Response in Disasters) project, which is funded by NIH's National Library of Medicine.

The WIISARD project allowed the San Diego Metropolitan Medical Strike Team to bring together scientists and engineers from the California Institute for Telecommunications and Information Technology with local and state police, SWAT, fire, haz-mat, and other first responders. In a simulation in 2005, the team was able to test the prototype of a video system that allows medical personnel to view a three-dimensional virtual environment generated by a live video stream.

In another new technology demonstration by the WIISARD project, first responders were provided wireless personal digital assistants, or PDAs, outfitted with software to help them keep track of victims' locations and triage status, capturing important medical data at the point of triage and transmitting it immediately back to hospitals and a command center using a Wi-Fi network. According to Bill Griswold, professor of computer science and engineering at the Jacobs School of Engineering, San Diego's Metropolitan Medical Strike Team "has realized that law enforcement is an integral part of medical disaster response, and to better coordinate that, they anticipate that technologies like this can be useful in communicating from law enforcement to medical responders without distracting law enforcement from their duties."

Griswold adds that "we've also had some interest from SWAT officials because these technologies would allow SWAT teams to communicate information silently back to their commanders. Currently they have to use hand signals or radios, both of which put them at risk from exposing their positions." Continued NIH funding to support this work is critical in keeping the nation safe and prepared for emergencies.

Similarly, but on a national scale, the CDC is an essential healthcare institution in emergencies, particularly in an age of biological weapons and biohazards that can spread as a result of natural disasters. Whether it is containing the threat of anthrax or limiting the spread of waterborne human disease, it is essential for the CDC to have effective communications capability in the first hours of an emergency.

Educational Institutions

In 1957 America rested assured of its status as a singular world power, convinced of its superiority on every front after the victory of World War II, the development and detonation of an atom bomb, and the resurgence of the economy that followed the Great Depression and allowed the United States to contribute to the rebuilding of Europe.

America could finally rest, and rest easy. And then, in October of that year, America's rest was rudely interrupted by Sputnik.

The Soviet Union's launch of an orbiting satellite haunted the American dreamscape with the sudden threat of communist missiles raining down from the skies, which sent schoolchildren under their desks to duck for cover. The director of development for the Army Ballistic Missile Agency at the time, German rocket scientist Werner von Braun, testified before a subcommittee of the House Committee on Education and Labor:

> Modern defense programs . . . are the most complex and costly, I suppose, in the history of man. Their development involves all the physical sciences, the most advanced technology, abstruse mathematics and new levels of industrial engineering and production. This . . . require[s] a new kind of soldier, who may one day be memorialized as the man with the slide rule. . . . It is vital to the national interest that we increase the output of scientific and technical personnel.

Sputnik's wake-up call led directly to the establishment of the Defense Advanced Research Projects Agency, or DARPA, which is credited with inventing the Internet. It also led directly to the passage of the 1958 National Defense Education Act. The NDEA allocated approximately $1 billion to support research and education in the sciences through 1962. The connection between education and defense could not be clearer.

Of course, educational institutions must have robust communications systems to warn and protect teachers and students. But to focus solely on American schools because they might be targets holding our children, our most valuable assets, would be to miss the lessons of the past. Our students and teachers, whether at the elementary or at the graduate-school level, must have the most advanced information technologies available if we are to develop the minds we will need to protect ourselves and find solutions to the various complex challenges in an increasingly complex world. To deny this access because a government investment may challenge the interests of private corporations misses the larger point that not doing so will rob those corporations of the very minds they need to stay competitive. To deny access to this technology may rob the nation of the resources it needs to save itself.

The importance of making advanced communications technology available to schools and students has been the subject of hundreds of reports over the past fifty years. Information-technology leaders in higher education were actively engaged in planning and deploying the networks that led to the formation of what many think of as the original Internet, the NSFNET of the late 1980s, along with successful efforts to generate congressional support for scientific and academic networks, leading to the High Performance Computing Act of 1991, and the National LambdaRail effort to build an all-optical, facilities-based network for leading-edge science and research.

The value of advanced broadband infrastructure is apparent in fields such as astronomy and genomics, but e-learning has barely scratched the surface of its potential. Students, particularly those who are not living at school, continue to have difficulty accessing broadband service. Undeterred, conservatives in the telecommunications industry continue to attack the Universal Service Fund (USF) program established by the 1996 Telecommunications Act, and have sought to undermine its effectiveness since its inception.

Yet the effectiveness of this program is undeniable. In 1998, at the beginning of the implementation of the USF program, only 14 percent of public school instructional classrooms were connected to the Internet; as of 2003, classroom Internet access was at 93 percent.

Nearly all public-library outlets today are now able to offer some Internet access. Yet in each funding year since 1998, requests for E-Rate discounts vastly exceeded the $2.25 billion available. Despite the clear need for and success of universal service, the Bush appointees at the FCC repeatedly undermined support for the fund by excluding cable companies providing advanced telecommunications services from the requirement of a universal-service contribution. Policy decisions like these only interfere with the goal of giving all Americans access to advanced telecommunications services whether they are poor, living in high-cost rural or urban areas, or living on fixed incomes.

Citizens remain our first line of defense and response in a natural disaster. If Americans are not connected, deployment will make little difference. USF support for advanced telecommunications services are clearly needed if all Americans are to be connected. A renewed commitment and a national broadband policy that puts universal access at the top of the list are past due.

The United States needs to move forward in a coherent fashion to deploy advanced telecommunications infrastructure, but not because we want to be no. 1. We have vulnerabilities at home that need to be addressed with some urgency. The possibilities resulting from the synthesis of powerful networks, computers, and databases have been the subject of a variety of blue ribbon panels, most notably the U.S. National Science Foundation report on cyberinfrastructure in 2003. More than five years later, another panel is called for, to advise a new administration and a new Congress.

The first work of such a panel should be to get accurate information on the deployment and capability of the various communications networks now operating in the United States, with a focus on the national goals of improved capacity (symmetrical speeds of ten gigabytes per second), redundancy, interoperability, and network neutrality.

We have a wide range of technologies available to communicate effectively. We should not choose between or among satellite broadband, Wi-Fi and Wi-Max, wireless broadband, power-line communications, and optical-fiber networks—all of these technologies should be invested in, along with new developing technologies to protect our defense and emergency needs at home. Because our citizens are our first line of defense or response, we need to make a commitment to universal service regarding advanced telecommunications services for all Americans.

As President Eisenhower said in 1955, "Our nation is sustained by free communication of thought and by easy transportation of people and goods." Our dependence on communications systems makes them more critical now than ever before. And as we pulled together and committed to the development of highways, satellites, and schools to win the Cold War, we must pull together now.

GETTING GOVERNMENT TO WORK AGAIN

"Doesn't anybody deserve a government that works?" has become the battle cry of at least one pundit. The answer is, of course, everybody does. But in order to obtain such a government, it helps to have trust in government. It also helps to acknowledge that, in a complex modern society, government is necessary. Note that we did not say a necessary evil—a knee-jerk sentiment that helped to get us in the fix we are in today. The idea that government is an inherent evil is, in fact, a misreading of classical economics and an exaggerated version of a market-based political philosophy, one that seems attractive in the seminar room but doesn't wash on Main Street or in the neighborhood.

Rather than assuming that government is the enemy, let's assume that, like all human projects, it will always be imperfect. Fortunately, that means it also can always be improved. In this section we offer some examples of how systematic reforms and new technologies can help us make government work better.

Outsourcing
Governance

I N THE YEARS SINCE 9/11, AMERICANS HAVE BECOME EVER MORE AWARE that contractors are doing much of the public's basic work—including a big share of sensitive, national security—related work—and in many cases at no small expense to taxpayers. In Iraq, Halliburton tends to the mess halls, Blackwater to armed security details, CACI International to Abu Ghraib prison. On the homefront, the Coast Guard bet its future on a multi-billion-dollar contract with Lockheed known as Deepwater Port to develop a new border-patrol system. The Department of Homeland Security watched a $100 million contract to hire baggage checkers balloon to $700 million. And the Federal Bureau of Investigation's $100 million-plus contract with SAIC to bring FBI case-information management into the twenty-first century came closer to bringing the agency to a halt.

When and how did the U.S. government become so dependent on private entities to do its public work? Some blame the Bush administration; others trace these policies to a Reagan- and Thatcher-era distrust for big government. In fact, the reliance on contractors to do the government's basic work is neither new nor an accident. It is the predictable and predicted outcome of a mid-twentieth-century bipartisan decision to grow government through the use of contractors. What's more, this decision was made in the name of science and was seen at the time by the science-policy elite as a progressive reform of truly constitutional dimensions.

This carefully managed reform yielded civilization-shaking developments: the Manhattan and Apollo projects, innumerable breakthroughs in biomedical research, and, of course, the Internet. But, as is now apparent from the day's front-page headlines, the twenty-first-century legacy of this reform is a government dependent on contractors to do some of its most important work, such as feeding our soldiers, protecting our diplomats, and collecting intelligence on the battle-field—often with too little official control or even awareness.

To begin to understand how we reached the point where dedicated contracts for specific projects with profound national-security implications morphed into a default determination to deploy contractors instead of civil servants as new governmental challenges arose, we must start with an understanding of science policy as it stood at the middle of the twentieth century.

From the Manhattan Project to the Military-Industrial Complex

America entered World War II with a small government and a long-standing preference to keep it that way. The war, however, required rapid deployment of the nation's scientific and industrial resources. The Manhattan Project to develop the first atomic bomb and other wartime research sponsored by the famed Office of Scientific Research and Development showed that the genius of America lay not only in its scientists and inventors, but also in its managerial talent—the ability to use the contract to organize private enterprise to public tasks of enormous complexity.

Thus, the atomic bomb was built in secrecy and speed by a combination of universities and industrial giants working on contract at secret, government-owned, contractor-operated sites throughout America. As Office of Scientific Research and Development historian Walter A. Macdougall observed, OSRD established the practice of "state funded but privately executed R & D. In a matter of minutes, patterns that had characterized American research throughout its history were undone."

Scientists, with their world-shattering public-private research-and-development projects, were providing the elixir that would make the world safe for a new kind of government.

That wartime success of the public-private partnerships led OSRD director and presidential science adviser Vannevar Bush to recommend to President Franklin Roosevelt, in the classic report *Science: The Endless Frontier*, that the government continue contract relationships with nonfederal organizations after the war. Incomprehensible today, but plausible in the shadow of mid-century totalitarianism, Bush needed to explain to corporate colleagues why taking money from Washington was okay. As Bush's memoir explains, his good friend, the president of Bell Labs, "was sure that we were inviting federal control of colleges and universities, and of industry for that matter, that this was an entering wedge for some kind of socialistic state."

Two decades later, in his 1965 book *The Scientific Estate*, science-policy eminence Don Price, first dean of the Kennedy School of Government at Harvard University, provided a classic explanation for developments calculated to satisfy their corporate and science beneficiaries. The United States, Price argued, needed more government to prepare to fight the Soviet Union, develop infrastructure, provide social welfare, and cure disease. The use of private contractors would permit the federal government to draw on private expertise, provide corporations with funding that would allay corporate fears that America was turning socialist, and would provide a force to countervail against the dead hand of a central official bureaucracy.

Price hailed the transformational import of the "fusion of economic and political power" and the "diffusion of sovereignty," both chapter headings in his book. Specifically, he argued:

> The general effect of this new system is clear; the fusion of economic and political power has been accompanied by the diffusion of sovereignty. This has destroyed the notion that the future growth of the functions and expenditures of governments would necessarily take the form of a vast bureaucracy.

By then, however, less flattering perspectives were already emerging. Most famously, in his 1961 farewell address, President Dwight Eisenhower declared the new public-private contract-based R&D partnerships to be the "military-industrial complex," an entity with "grave implications" for "our liberties [and] democratic processes."

Early in his presidency, John Kennedy waded into this debate by commissioning a cabinet-level report to examine the implications of the contract-dependent R & D. The 1962 Bell report (after Budget Bureau Director David Bell) served as a springboard for the first public congressional hearings on Rand and other government-created contract think tanks and systems managers that had been spawned at the nexus of government and the aerospace industry. It is the last, best, and indeed only (save for Ike's speech) White House review of the wisdom of government by contract, and foretold much of what has since come to pass.

The report declared that the "blurring of the boundaries between public and private" raised profound questions about the axiom that officials must be able to account for the work of government. In the short term, the report said, the use of contractors to respond to Cold War emergencies made sense. But over the longer term, it warned, the axiom of official government control over government contractors would be challenged unless corrective action were taken.

The Bell report put its finger on the problem: the disparity between the rules of law governing officials and contractors. Americans, ever concerned with big government, have enacted during the past two centuries countless laws to protect against official abuse. These began with the Constitution and its Bill of Rights, which define the limits of government and provide for individual rights against government abuse, and now include laws on ethics, pay, freedom of information, and political activity.

By and large these laws do not apply to those outside government— even contractors doing government work on taxpayer dollars. They do not apply on the dual premise that officials will have the capacity to oversee contractors, and that the qualities for which contractors are valued may be constrained if they are subject to rules governing officials. Yet, as the Bell report prophesied, when the decision to rely increasingly on contractors for key government work was coupled with the freedom of contractors from rules limiting government officials, the predictable effect would be the erosion of federal oversight. It also promised to undermine government efforts to recruit talent.

Why, the Bell report asked, should experienced government officials choose to remain in government service when they can work as contract employees who are not governed by official pay caps and the stringent constraints of official ethics rules—and do work no less interesting than that in government?

But the Bell report backed away from answering the basic questions it raised. The new public-private mix, it found, was essential to Cold War programs. "Philosophical issues," the report said, needed to be deferred to a later date. Thereafter, as government grew, third-party government grew along with it.

From the Bell Report to Abu Ghraib

Cold War agencies such as the Atomic Energy Commission, the Department of Defense, the National Aeronautics and Space Administration, and the U.S. Agency for International Development, provided the initial template for the deployment of contractors as a permanent work force for the performance of core public tasks. From the get-go, and in its present incarnation as the Department of Energy, the nation's nuclear-weapons complex has been essentially government-owned and contractor-operated.

"NASA," the *Washington Post* observed following the space shuttle *Columbia* tragedy, "may hire the astronauts," but "at the Johnson Space Center, the contractors are in charge of training the crew and drawing up flight plans. The contractors also dominate mission control, though the flight directors and the 'capcom' who talk to the crew in space are NASA employees."

In the 1960s and 1970s, these contracting models were transferred from Cold War agencies to civilian agencies as the Great Society programs initiated by President Lyndon Johnson rushed to embrace contractor-generated management products, such as PERT (Program Evaluation and Review Technique) and PPBS (Planning, Programming, and Budgeting System) incentive contracting, and systems analysis. With the birth of each new agency or program, contractors were trained and deployed to do the government's basic work. Over time, the United States became home to a massive contract work force that promised to solve a vast range of problems, from the inner cities to the jungles of Vietnam.

In 1971, looking back on developments, John Corson, the famed Washington economist, management consultant, and international adviser, marveled at what had been wrought. "There is," he wrote in the book *Business in the Humane Society*, "little awareness of the extent to which traditional institutions, business, government, universities, and others have been adapted and knit together in a politico-economic system which differs conspicuously from the venerated patterns of the past." Postwar contracting, he said, was a "new form of federalism."

From the perspective of the Nixon White House, to be sure, what had been invented was not a new way of government, but a new way of patronage. The *Political Personnel Manual*, uncovered by the Senate Watergate investigations, showed that Nixon White House staffers, seeking to catalogue ways to control the presumably democratic-leaning civil service, were bemused to discover that, under the guise of efficiency, JFK and LBJ had used cutbacks in the civil service to hire friends as contractors—with, the Nixon staffers noted, higher taxpayer costs. Rather than dismantle the system, however, the Nixon administration put it to work for its own ends. At the Office of Equal Opportunity, the central war-on-poverty agency, two young administration officials named Rumsfeld and Cheney turned the tables and hired their own contractor, Booz Allen, to help take control of the agency, shuffling out liberal-leaning civil servants under the guise of agency reorganization.

In the 1980s and 1990s a new generation of reformers—the privatizers, downsizers, and reinventors—came to argue for the reform of big government, often with little evident knowledge of the history or legacy of the reform that had already been going on for decades in America. When, in 1993, the Clinton administration announced its intention to "reinvent government," it declared it would reduce big government by further reducing the numbers of civil servants, with little demonstrable recognition that this had long been the means for growing, not reducing, government.

Reinventing government followed in the footsteps of the Price logic—the diffusion of sovereignty was a good thing that could be kept under control by modern management techniques such as performance contracting. President George W. Bush entered the White House committed to a similar philosophy. One of the pillars of the Bush management agenda was to put out for competition up to 850,000 civil service jobs. And after 9/11, the new Department of Homeland Security was launched in large part on the wings of a contractor work force, with limited official oversight capacity.

The overall task was to cut red tape and make the use of contractors simpler and more efficient by, for example, centralizing purchasing through the General Services Administration so agencies could essentially buy from a catalogue without need for further competition. As Iraq showed, however, an effect of the reform was to maximize potential for unaccountability. When the public learned that a contractor called CACI had been at work interrogating prisoners at Abu Ghraib, there was a scramble to find out how the Army hired the contractor. It turned out the contract was originally awarded by the GSA, assigned to the Department of the Interior for administration, and then used by the Army in Iraq. Apparently unbeknownst to the GSA, the Army used the contractor for purposes that were (unlawfully) beyond what the GSA contract provided for—indeed, for purposes contrary to the Army's own prohibition against the use of contractors to perform such intelligence gathering and contrary to the rule against using contractors to serve, in essence, as integrated members of the government work force.

Even before the war began, in March 2002, the secretary of the Army sent a troubled memo to top Pentagon officials confessing that Pentagon Army planners had little clue as to the size of the contractor work force, the costs associated with it, or "the organizations and missions supported by them." The next month, the Army told Congress that its civilian-support work force was clearly large but that its size was uncertain, with the Army's own estimates varying from 100,000 to 600,000. In short, assuming the Iraq war were to be waged, contractors would be a major component—and inability to oversee them effectively a predictable consequence.

From Today's Headlines to a Re-envisioning of Twentieth-Century Reform: Alternative Visions

The domestic "diffusion of sovereignty" that Price foresaw, and indeed urged, has now meshed with a larger, global diffusion of sovereignty. Research and development is today performed by networks of

multinational corporations, global universities, and numerous governments. That dynamic, still young, may give rise to its own global problems.
But here at home Price's vision has—as the Bell report prophesied—
left a legacy that unquestionably demands attention. We can no longer
assume that the official government work force has the capacity to
understand, much less control, the basic work of government.

In decades past, either of two alternative frameworks were said to
ensure ongoing federal control over the federal contracting work force.
Both have proven inadequate to the task as the balance of power has
increasingly shifted away from government and to the private sector.

First, law and policy enshrine what might be called the "presumption of regularity/rule of law" framework. The presumption of regularity is a long-standing legal tradition that says officials can be presumed
to act with the good faith, diligence, and competence expected of them.
Under the Western rule-of-law tradition, the presumption of regularity
envisions that, such assumptions notwithstanding, officials must be
subject to rules that define and limit government authority and protect
against government abuse. But as long as the presumption of regularity is valid, then these rules for officials need not apply to contractors
because those contractors will be accountable to public officials.

The problem, as the front pages increasingly tell us, is that this presumption of regularity does not reflect reality where contractors have
been engaged to do the government's basic work but are not subject to
laws and policies governing civil servants. When it became apparent,
at the dawn of contract reform, that the presumption of regularity was
imperiled, the Eisenhower White House issued a policy declaring that
only officials can perform "inherently governmental" functions. That
policy has been dutifully reiterated by every White House since—most
recently by the Bush administration, virtually coincident with the start
of the Iraq war and the wholesale deployment of contractors on the battlefield. Today the policy is but a fiction or fig leaf, hiding a very different reality.

The second framework might be called the "governance/account-
ability" perspective. This vision, a modern-day version of Price's
diffusion-of-sovereignty idea, holds that the work of the public is best
done by a mix of government, civil society, and market organizations.

Attention is focused on results and not on the "boundary between public and private," as long as there is accountability for that work.

In the world of governance, accountability is to be provided in three interlocking ways:

- Modern management and social-science techniques, which will align public and third-party interests by properly structuring contractual performance standards and incentives.
- The force of competition between or among contractors and interest groups, which will help keep the system honest.
- Transparency as an aid to the first two tools.

While not forsaking the premise that officials must be able to account for all government work, proponents of this governance and accountability model suggest that the civil service work force must be transformed into a work force that functions substantially, or even primarily, to manage third parties. The problem, as the front pages once again tell us, is that these tools are no longer sufficient, and can even be counterproductive, in this age of diminished official oversight.

Thus, though competition in procurement is a key tenet of procurement law today, the reality is that competition is often limited because skilled or eligible applicants may be few (particularly in security areas where work forces must have clearances), and because for many such projects, the scale is so large that to even make a proposal is for all but a few companies prohibitively costly. Similarly, performance measures are hard to come by for novel or highly specialized public tasks, and enforcement may require extraordinary resources, undermining efforts to penalize poor performance. Because the work has to get done, and because too harsh a penalty risks eliminating a competitor without whom there may be no competition at all, performance standards are often allowed to slip. As for transparency, it is nice in theory, but the contract work force too often remains largely invisible within agencies and all but unknown to the public at large.

As a consequence of the limits of these two frameworks, and in the absence of coherent congressional and executive oversight, the rules of the game as it is played today owe more to a third model, which might best be called the "muddling through/common law" model. Under this model, when crises arrive, they are handled with limited regard for the big picture.

The model draws upon a long-standing Anglo-American legal tradition, reflected in the way public utilities are regulated, which accepts that rules of public law should apply to those who perform public tasks, even if those doing the tasks are nongovernmental actors. But it applies the rules on a piecemeal basis.

Yale University professor Charles Lindblom's classic article *The Science of Muddling Through* explained that such a process is in many regards the American way; but it has its weaknesses, as the evolution of reform by contract well illustrates. First, successful muddling through requires a healthy crew of diverse and well-funded nongovernmental watchdogs to keep policy and practice honest. Government contracting is an insider's game. Unfortunately, public watchdogs have been few and far between.

Without such oversight, the muddling through approach runs a serious risk of focusing too much on individual situations and not enough on larger, troubling trends. Consider the history of conflict-of-interest rules for contractors. In the mid-twentieth century, no such rules were in place. As Price observed, during the 1950s "no Congressmen chose to make political capital out of an investigation of the interlocking structure of corporate and government interests in the field of research and development." Indeed, the concept of "organizational conflict of interest," or OCI, was conceived and invoked only when some contractors felt that other contractors were using their inside track to unfair competitive advantage. At the time, it left wholly unaddressed those circumstances in which collective contractor interest was served but public interest disserved.

In the late 1970s, the notion of public interest was added to the OCI concept. Even so, contractors remain eligible to be hired even when conflicts exist. Moreover, while civil servants can go to jail if they work for, say, the Department of Transportation and General Motors, no criminal conflict-of-interest rules govern private contractors. Worse, in the private sector, potential conflicts are exempt from disclosure under the Freedom of Information Act, and there is no routine independent audit of the conflict-review process.

Most problematic, as Lindblom explained, the model of muddling through produces change by increment that serves the interests of those pushing for those changes but generally with little regard for the big picture, which may become increasingly off kilter. In the case of government contracting this has led to the gradual erosion of official oversight capacity and, with regard to accountability, a disconnect between law and reality.

Reform in Need of Revision

The time is long overdue for truth-in-government budgets, revelatory organization charts, and basic data on the contract work force. The White House and Congress and the public all deserve an honest view of the work force that is doing the basic work of government. Among the pressing questions that need to be answered:

- How many government functions vital to national security and well-being are now being, or are in danger of being, contracted beyond official oversight capability?
- In the awarding of contracts to perform the work of government, are there competitive markets? Can competition be relied on where official oversight is limited? Can performance measures be relied on as used and are they useful?
- Where contractors work for multiple agencies and multiple components within agencies, and where they boast work forces that oscillate between government and private sectors, how well do they perform the important role of networking, and what conflict-of-interest problems are presented?
- With consideration of the role of contractors and other third parties, including whistle-blowers, in contractor oversight, what is the scope and adequacy of the procurement-oversight work force, and how should its organization be shaped for the future?

Finally, there must be an effort to harmonize the law of public service with the current reality, and to consider the possibility of a broader, new ethic of public service. Today, dual sets of laws and policies govern contractors who may work side by side with public servants performing the same work. That means the laws enacted over two centuries to define and limit government, and to protect Americans against government abuse, increasingly do not apply to those doing the real work of government.

There must be public review and comparison of the differing rules that apply to federal employees and to nongovernmental actors in the performance of the government's work, including those rules governing constitutional protections afforded citizens in relation to official conduct; political activity; ethics; and transparency. As became clear from revelations about the security contractor Blackwater and its operations in Iraq, the review must include relevant international, as well as domestic, rules.

This task may not produce clear-cut answers and will certainly not be a panacea. Simplistic application of uniform rules to public officials and private contractors may well be counterproductive, negating specific and complementary qualities for which contractors and civil servants are valued. Moreover, the procurement system has its own complex body of laws, rules, and policies, and any attempt to harmonize procurement and personnel law will be difficult. But that does not preclude the possibility of developing generally acceptable ethics principles—an ethics of public service—applicable to all those paid by taxpayers to serve the public.

The need for an ethics of public service comes more into focus when we consider that the expertise and experience needed to manage contractors increasingly lies in the contractor work force itself, creating an "information asymmetry" that society has traditionally sought to balance through the promulgation of special ethics rules for experts who have special knowledge. Perhaps most famously, doctors and lawyers are obliged to consider client/patient interests independent of their own bottom-line profit. But the need for special principles to govern private contract experts who perform public service has been obscured by the presumption of regularity. In contrast to a patient or a legal client, the U.S. government is presumed to have the resources—the authority, the people, the knowledge, and the money—to make decisions and protect itself, a presumption given legal form in the presumption of regularity and the related inherently governmental policy.

An ethics of public service that applies to all those who do the work of government could go a long way toward resolving this dilemma. The aim should be to identify and address problems raised by the combined effect of the information asymmetry that characterizes expert/ client relations elsewhere and the mismatch between current laws and policies—which assert and presume regularity—and the reality that officials are no longer fully in control. If the revolving door between government service and private profit, which has provided many contractor-officials with an inside understanding of government, has led to an erosion of the presumption of regularity, it might also become a force to help restore this presumption with the addition of an ethics of public service.

The United States may be unique among modern governance systems in its scope of reliance on contractors, but the approach appears to be growing in global popularity. As others turn to the U.S. system as a model for study and possible adoption, America should recall the ingenuity that was at the heart of its mid-twentieth-century reform and apply that same creative spirit to solving the problems that are the legacy of that reform.

Fixing Our Fractured Food-Safety System

THE AVERAGE CARLOAD OF AMERICANS PULLING INTO A REST AREA FOR lunch probably feels reasonably confident that the U.S. government is up to the task of ensuring that a bite to eat won't spell a quick end to their road trip. But as things stand, responsibility for the safety of even the simplest of meals falls messily to a wide array of federal agencies and offices that together make up our fractured food-safety system.

As the Government Accountability Office (GAO) has dryly noted, an open-faced ham sandwich sold at a highway rest stop is the responsibility of the U.S. Department of Agriculture and subject to daily inspections. But add a second slice of bread and it becomes the Food and Drug Administration's job to check in on the sandwich, which it does about once every five years.

And so it goes with just about every type of food Americans consume, resulting in a food-safety system that verges on the absurd. The FDA regulates chicken broth, but beef broth is under the USDA's watchful eye—except for dried soups, in which case the agencies swap duties. Responsibility for packaged baked beans depends on whether the meat in the can is pork chunks (FDA) or bacon (USDA). And under which agency's purview a pizza falls depends on whether it is of the cheese-lover, meat-lover, or seafood-lover variety.

The American food-supply chain's remarkable ability to assemble a coherent meal independent of the season or the geographic sources of ingredients masks our regulators' stunningly uneven oversight of the production and assembly process. A single cheeseburger purchased at the to-go window of a fast-food chain alongside any highway in America can contain a beef patty made from a hundred heads of cattle, cheese from the milk of a dozen dairy farms, lettuce from Arizona engineered to look fresh for days on end, and tomatoes "strip-mined in Texas," as Garrison Keillor once joked. Rivaling this supply-side patchwork, the inspection of these ingredients rests with a bureaucratic alphabet soup of agencies. Take the 2004 case involving two USDA agencies, the Animal and Plant Health Inspection Service and the Food Safety and Inspection Service. At the very same time APHIS was testing the carcass of a cow infected with bovine spongiform encephalopathy, or mad-cow disease, FSIS was clearing its beef to head out to market.

Food production and distribution today is a well-oiled system, as we saw in 2006 when spinach contaminated with *E. coli* spread quickly from California's Salinas Valley out across the country, turning up from Oregon to West Virginia. But if the food supply is a symphony of cooperation, the federal oversight of that system is a nearly completely atonal chorus. More than a dozen federal agencies have differing and overlapping areas of responsibilities and ways of doing business.

Plans for a single government public-health body to take command of food safety have long foundered in Congress, but interest has been growing as the news has become filled with tales of outbreaks, from multimillion- pound beef recalls, to salmonella-tainted peanut butter and pot pies, to melamine-laced imports from China. According to the CDC, each year in the United States an estimated 76 million illnesses and 5,000 deaths are caused by unsafe food.

Our fractured federal system for overseeing what we put into our mouths—a system crafted for the days when Upton Sinclair's 1906 book *The Jungle* had turned Republican President Theodore Roosevelt into the nation's leading progressive food reformer—is a big part of the problem. Since the days of *The Jungle*, the U.S. food-safety system has evolved in fits and starts. When President Franklin Roosevelt sponsored the creation of the FDA in the 1930s in response to food and drug scares, much responsibility was left behind at the USDA.

Visit the food section of Recalls.gov, a joint project of several federal agencies, and the problem immediately becomes clear. To go any further on the Web site, you must click the logo of either the FDA or USDA—leaving it to eaters to know which federal agency is responsible for overseeing the safety of which foods. What's more, recall notices on a government Web site offer a conservative measure of the problem and a false sense that matters are under control, given that food recalls are nearly always voluntary.

Taken together, the American food-safety system is one designed to put out fires, not to ensure food safety in line with modern science. "When we had the spinach episode, everyone acted like it was a great surprise," former FDA Commissioner Lester Crawford, a Bush appointee and long-time federal food-safety official, told the Center for American Progress. "But the likelihood of something bad happening [with the food supply] is always quite high." Asked about the working relationship between FDA and USDA officials, Crawford said they "generally don't bump into each other. I don't know if I ever tried to make a phone call to the USDA. And if I did, I don't know if it would have been returned."

Marion Nestle, New York University professor and the author of the book *Safe Food,* who has served in a number of food positions on the federal level, has been equally blunt about the U.S. food-safety system, calling it "an overlapping system with huge gaps where everybody blames everybody else."

The GAO, which has long called for a single food agency, recently declared the current system "high-risk." When it comes to the FDA, says Lisa Shames, GAO's director of food safety and agriculture issues, part of the problem is "a mismatch between funding and food oversight responsibility." Specifically, the FDA oversees four-fifths of the food supply but receives just one-fifth of the total federal budget for the effort. The fiscal year 2009 presidential budget called for increasing the FDA's long-stagnant funding level to $662 million, a meager 7 percent boost covering little more than inflation. The FDA itself says it needs an additional $275 million just to beef up its overseas inspections.

Beyond the FDA's meager budget is the challenge of having an agency with so vast and diverse a mission, one responsible for the safety of America's food and drug supply. Said Crawford, "I just can't recall an incident when I said, 'My gosh, thank God we have the drug people with the food safety people.'" Crawford discounts the possibility of finding agency-level leadership equally skilled in food science and pharmaceuticals. "They just don't make people like that," he said.

The USDA, home to the majority of agencies with food-oversight duties, is a different entity with its own special challenges. In 2003, then-Secretary Ann Veneman lamented that the department was bound by laws that predated the Model T. Mike Doyle, director of the University of Georgia's Center for Food Safety, said the USDA "is in the plant to look at gross morphology, basically looking for lesions" on animals heading for slaughter. "That was relevant back when the statutes were written," Doyle said—in the turn-of-the-century days when Sinclair wrote of meat in Chicago's packinghouses being "moldy and white, stinking and full of maggots." But visible problems like rotting meat aren't today's concerns, Doyle said. Today's worry: bacterial pathogens such as *E. coli* and salmonella, both of which are invisible to the naked eye.

What's more, the USDA is a department internally conflicted. Its primary role in Washington is to promote the food trade—to boost the amount of American pork the Chinese eat, not to worry over whether the pork Americans consume is safe to eat. The GAO recently profiled seven countries (Canada, Denmark, Germany, Ireland, the Netherlands, New Zealand, and the United Kingdom) that have successfully consolidated food-safety oversight under one roof. Most interesting is the holistic farm-to-fork approach of European Union countries. Ireland is a typical case, moving its food-safety agency under the auspices of its existing public-health authority, in recognition of the fact that the *raison d'etre* of its own Department of Agriculture is promotion, not policing.

Added to America's food-safety challenge is the fact that we're now pulling an enormous amount of food into our supply stream from overseas—up to 15 percent of what we eat, by volume—and inspecting a minuscule 1 percent of it at most. The current regime sends a message to food producers in the wake of the melamine scare, said NYU's Nestle. "The Chinese were very frank about it," she said. "'You asked us to give it to you at the cheapest prices. You didn't say anything about quality.'" Even occasional point-of-entry inspections could act as a deterrent. But today's nearly non-existent inspections simply set the expectation that the fractured U.S. food supply is willing to absorb foods of dubious quality.

What's the solution? For years now, diverse voices in Washington—from the GAO to the National Academies' Institute of Medicine and National Research Council to former Department of Homeland Security Secretary Tom Ridge—have been calling for the creation of a single food-safety agency, a player in the federal bureaucracy with the necessary mission, might, and budget to ensure a safe food supply. On Capitol Hill, Senator Richard Durbin (D-IL) and Representative Rosa DeLauro (D-CT) have repeatedly introduced the Safe Food Act, a legislative vehicle that would not only create a Food Safety Administration but also establish a firm schedule for inspections and give the new body the power to invoke mandatory recalls.

Interestingly, given Ridge's past support for the idea, GAO recently eased its strong call for single-agency plans, citing the difficulties that arose in the creation and amalgamation of the Department of Homeland Security. A spokesperson for DeLauro counters GAO's concerns by arguing that the creation of DHS was a different effort entirely —an attempt at unifying offices and agencies with unique aims and cultures. She argues that the creation of a Food Safety Administration would be more akin to federal reorganizations like the 1947 establishment of the Department of Defense, which united federal agencies and offices already committed to a common mission.

Even industry opponents of current single-agency proposals, such as the National Cattlemen's Beef Association, are quick to say they are united behind the common goal of ensuring a safe food supply. The Cattlemen's Phyllis Marquitz has said she objects to Durbin and DeLauro's plans as "political solutions that rearrange agency structures but do little to show potential for real-world practical change." But the beef-industry spokeswoman adds that she judges beef producers to be receptive to a convincing case that one unified federal food-safety agency is indeed the best way to ensure safe food.

Former FDA Commissioner Crawford echoes the sentiment. In his experience, everyone involved in the food chain "agree[s] with the idea that we have to put safety first," he said. "The question is how we get there."

Of course, restructuring the way the federal government handles food safety is no easy task. Agency heads are generally loath to give up jurisdiction and budget. From deep-pocketed meat lobbyists to members of House and Senate agriculture committees, many in Washington with a role in the food-supply chain have an interest in maintaining the idea that food safety is an industry issue rather than a public-health concern.

But perhaps most important is Congress's limited supply of attention. It's been nearly seventy years since the public last rose up to demand safer food, forcing politicians to act. Given the challenges posed by our increasingly globalized food supply and the mounting reports of food-borne outbreaks, the time seems ripe for a new wave of food-safety activism and a real commitment from Congress to address the issue.

Stress Test: Turning to Technology for Those Bridges to "No Wear"

S OMETIMES IT TAKES A DISASTER TO BRING A LONG-SIMMERING PROBLEM to light. Such was the case on August 1, 2007, when an eight-lane interstate bridge in Minneapolis collapsed during evening rush hour, killing 13 people and injuring 144. The collapse, and the failure to anticipate it, called into question the adequacy of current bridge-inspection methods. Why were problems with the bridge not identified? Could similar problems be missed elsewhere? Might this happen again?

There is good reason to worry. Before it collapsed, the Minneapolis bridge was one of more than 70,000 nationwide declared by the Department of Transportation to be structurally deficient. One in three urban bridges fall into this category. Yet "structurally deficient" is a broad classification, and current inspection methods cannot be relied on to identify those bridges that are truly on the brink of collapse.

"We do not know which bridges should be taken out of the system, and which should be maintained," said A. Emin Aktan, a professor of civil engineering at Drexel University and director of the Intelligent Infrastructure and Transportation Safety Institute.

Answers may come from sensor technology. In September 2008, Minneapolis opened a new Interstate bridge that contains more than 300 embedded sensors to take detailed, real-time measurements of the bridge's condition. Unfortunately, this technology is being used on less than a handful of bridges nationwide.

The current system of conventional inspections is anything but high-tech. Every two years, each government-owned bridge is required to receive a routine inspection, in which technicians or engineers observe the bridge and take measurements of its physical condition. Underwater structures, meanwhile, must be inspected by divers every five years. There are guidelines but no requirements for in-depth inspections, which can include things like probing of the bridge, laboratory analysis of bridge material, and testing of surrounding environmental and water conditions.

This heavy reliance on visual inspection is inadequate for three major reasons. First, such inspections are susceptible to human error. Indeed, a 2001 study by the Federal Highway Administration found that inspectors regularly missed problems and inconsistently rated bridge conditions. Second, there are long intervals between required inspections, during which time serious problems may emerge. And third, even high-quality visual inspections are inherently superficial and can fall short of producing the detail necessary to spot deficiencies.

This is not to say that visual inspection is unimportant; visual inspection is crucial to assess bridge conditions, in particular to identify cracks and corrosion. But more is needed to assure the safety of the nation's bridges.

That's where sensor technology comes in. Instead of relying on sporadic and error-prone observations, matchbox-sized wireless sensors can be attached to or embedded within bridges to take precise, continuous measurements of virtually anything relevant to a bridge's condition, including strain, tilt, vibration, temperature, and seismic activity. This sort of data is particularly important as the nation's bridge population ages—the average U.S. bridge is now forty years old—and traffic and truckloads continue to increase, causing more rapid deterioration.

The Minneapolis collapse created a political opportunity to modernize bridge monitoring. In its aftermath, Secretary of Transportation Mary E. Peters initiated an ongoing review of the agency's bridge- inspection program to, in her words, "make sure that everything is being done to keep this kind of tragedy from occurring again."

Congress, too, got engaged. Rep. James Oberstar (D-MN), chairman of the House Transportation Committee, introduced legislation to significantly improve bridge-inspection requirements as part of "a data-driven performance-based approach to systematically address structurally deficient bridges on our nation's core highway network."

Sensor technology can help meet the goals expressed by Peters and Oberstar. What's needed now is a plan to move forward.

Recent progress in Connecticut could serve as an example. The Federal Highway Administration awarded funding to that state's Department of Transportation and the University of Connecticut to deploy and study different types of sensor systems for long-term bridge monitoring.

"The goal is to generate information between inspections, so that if there's a major change, we can take action to prevent something catastrophic from happening," said project leader John DeWolf, a professor of civil engineering who became involved in bridge monitoring following the 1983 collapse of the Mianus River Bridge on Interstate 95 in Greenwich, Connecticut.

Over the last several years, six bridges in Connecticut have been outfitted with unique sensor systems. Five of these are wired systems, in which cables connect the sensors to a computer. The sixth relies on solar-powered wireless sensors. This wireless system is particularly exciting because it holds great promise to be more widely duplicated across the country.

It can take a great deal of labor and expense to run cables over a bridge, especially one that is large and difficult to access. For a wireless system, however, cables are not an issue. Sensors merely need to be placed in desired locations on the bridge. Installation typically takes no more than a few hours, at a cost less than half that of a wired system.

Because of these advantages, DeWolf decided to go wireless for Connecticut's longest bridge, the Goldstar Bridge, which crosses the Thames River on Interstate 95 in New London. Like all new technologies, wireless sensors are expected to get much cheaper over time. But even now they are affordable. Installation of twelve sensors at the Goldstar cost about $30,000.

Over the long run, sensors may even pay for themselves by more precisely identifying when and where repairs are needed. Ten wireless sensors were recently used to test stress levels from passenger trains on the Ben Franklin Bridge, which crosses the Delaware River from Philadelphia to Camden, New Jersey. The state believed the bridge was in need of major repairs based on advice it received from an engineering consultant. But data gathered by the sensors showed the bridge was in fact secure, saving tens of thousands of dollars in unnecessary repairs.

Sensors can also reveal problems as they emerge, before there is visual evidence such as cracking. This allows remedial action to be taken in time to head off serious structural damage, which can be expensive to repair. "If you get to it quickly and fix it, it's not going to be a major problem," said Mike Robinson, vice president for sales and marketing at MicroStrain Inc., which developed the sensors for the Ben Franklin Bridge. "You can reduce the overall life-cycle cost of the bridge."

DeWolf approached MicroStrain to develop the solar-powered sensors specifically for the Goldstar. The sensors used on the Ben Franklin Bridge were powered by batteries—fine for short-term testing, but not long-term monitoring. Batteries eventually run out of power and need to be changed or recharged, which is difficult on a bridge like the Goldstar, where sensors are in hard-to-reach locations.

The solar-powered system relies on photovoltaic panels to harvest energy from the sun. These panels are connected to the sensors to supply power for daytime monitoring and to recharge batteries for overnight observation. This system is expected to generate power for years with little or no maintenance. MicroStrain is developing other solutions for long-term power, including mini-wind turbines and super-efficient battery-powered sensors, according to Robinson.

MicroStrain first installed its solar-powered system on the Corinth Canal Bridge in Greece to monitor seismic activity. There, the sun is strong enough for continuous monitoring, which is crucial given the unpredictability of seismic activity. At the Goldstar, where the sun is not as bright, data are gathered for five to ten minutes every hour to conserve energy. For what's being measured, strain and vibrations, this is considered plenty sufficient.

The data collected are temporarily stored on the sensors and then downloaded daily to an onsite laptop computer. From there, the data can be remotely accessed through a DSL connection. Of course, it is not possible to manually analyze the voluminous amounts of data generated. Instead, automated systems are programmed to comb through and pick out relevant information for DeWolf and his team to review.

Ultimately, this information can help confirm whether the bridge is safe. Vibrations, for example, can be monitored to ensure that they do not exceed potentially dangerous thresholds. For the vast majority of the nation's bridges, this sort of information is not available. Indeed, Connecticut is now the only state using sensors for long-term monitoring of multiple bridges. Other states rely on the same visual inspection methods that failed in Minneapolis.

"Let's not debate that visual inspection has proven insufficient," Aktan said. "Instead, we should focus on strengthening bridge monitoring, so that one day there will be little worry about another bridge collapsing."

But how should the nation scale up such efforts? The Federal Highway Administration, unfortunately, does not systematically identify priorities among the bridges it deems structurally deficient, nor are bridges of greatest concern necessarily given more attention. Rather, every bridge is subject to the same biannual requirement for visual inspection regardless of physical condition.

High-priority bridges need to be identified and should be repaired as quickly as possible, of course. But sensors should not only be installed on the very worst bridges. Ideally, they should be used to assist routine maintenance, too, so that bridges never get to the point of imminent collapse. Ultimately, we will need a system to smartly and economically deploy sensors to monitor the nation's entire bridge population.

The first step in this process is to classify bridges according to type. A suspension bridge like the Brooklyn Bridge obviously has different characteristics than a truss bridge like the Goldstar and the Interstate 35 bridge that collapsed in Minneapolis. But even bridges of the same general type can have critical differences. Truss bridges, for example, use a variety of bracing designs, may or may not use pins to connect joints, and may carry traffic on the top, middle, or bottom of the structure.

Bridges will deteriorate in different ways and at different rates depending on such variables. Currently, however, the nation's bridges are not carefully categorized by similar design features. This information is needed to determine which bridges to outfit with sensors.

Because similar bridges can be expected to perform alike, it is necessary to install sensors only on a sample from each category. Again, this sort of sampling is not part of the current monitoring system; instead, each bridge is subject to the same biannual inspection. "Looking at each bridge as an individual is ridiculous," Aktan said. "There are tremendous similarities between certain types of bridges, but we don't leverage knowledge about those similarities."

The Federal Highway Administration recently launched an initiative—the Long-Term Bridge Performance Program—that begins to move in this direction. The goal of the program is to generate "high-quality, quantitative performance data" based on a sample of the nation's bridges, likely numbering 500 to 1,000 bridges representing the majority of structure types. This includes data on deterioration and its causes—traffic load, corrosion, fatigue, and weather, among others—as well as the effectiveness of maintenance strategies.

As part of its data-gathering efforts, the highway administration intends to subject the bridges in its sample to detailed periodic evaluations, over at least a twenty-year period, using sensor technology and other state-of-the-art monitoring tools. In addition, a subset of bridges in the sample will be instrumented to permit continuous monitoring, while decommissioned bridges will undergo forensic autopsies.

Congress created this program under legislation enacted in August 2005, with funding authorized through fiscal year 2009. The highway administration requested $20 million a year but will have to operate with only about $5.5 million a year over the first four years. Thus, tough decisions will have to be made as to which parts of the program to launch immediately and which to postpone pending higher levels of funding.

The initiative will be especially valuable in determining what data to collect and what those data mean. In particular, it is often unclear what and where to measure. If sensors measure the wrong things or are placed in the wrong spots, they may miss critical deficiencies. The highway administration's research will begin to identify key factors and pressure points in the deterioration of different types of bridges. "A doctor knows where to take a patient's pulse," Aktan said. "We need to know where to take the bridge's pulse."

A critical part of this process is knowing how to interpret the pulse so sick bridges are diagnosed and treated. The vast majority of bridges lack baseline performance data—that is, data collected at the time they were built—from which to judge deterioration over time. Without this information, there is uncertainty about a bridge's optimal performance and exactly what constitutes poor performance.

The highway administration intends to address this problem by comparing newer and older bridges of similar type to identify and predict life-cycle changes. This should bring into sharper focus the large amounts of data generated by sensors. "The problem we have now is making sense of this data," said a government engineer involved in the Long-Term Bridge Performance Program. "That's what we are trying to address. Determining the sample of bridges is the most critical step." The agency has already developed methodology to identify bridges for the sample. A final selection of bridges will be made in collaboration with the Center for Advanced Infrastructure and Transportation at Rutgers University, which was contracted in April 2008 to oversee the program's day-to-day operations.

The highway administration's research deserves the full support of Congress and the president. The amount currently appropriated is barely enough to get the program off the ground. One enduring problem, unfortunately, is the tendency of Congress to fund transportation research through earmarks to specific universities or private contractors. These earmarks sometimes go to worthy projects, but frequently they are awarded according to political considerations rather than merit.

Moreover, because funding is disjointed and somewhat arbitrary, transportation research is not well integrated and coordinated. The Long-Term Bridge Performance Program can help add cohesion by drawing together information generated by disparate research efforts, including other highway administration–funded initiatives, such as the sensor project in Connecticut. "We will try to piggyback on other research projects and make them fit into this national approach," the agency engineer said.

It couldn't happen too soon. For years, the president and Congress have repeatedly deferred needed maintenance of bridges in favor of other budgetary priorities. This shortsightedness will cost the nation far more in the end, as the scale and severity of needed repairs balloon and become impossible to ignore. In the aftermath of the Minneapolis disaster, the Department of Transportation released $55 million in emergency funds, and Congress authorized $250 million for rebuilding.

Other critical infrastructure—including roads, dams, and levees—are similarly deteriorating and could also benefit from enhanced monitoring through sensors. Substandard road conditions contribute to 30 percent of all fatal highway accidents, according to the highway administration. More than 3,500 dams are unsafe or deficient, many of which may not hold during significant flooding or an earthquake, according to state inspectors. And nearly 150 of the nation's levees pose a high risk of failing during major flooding, according to the U.S. Army Corps of Engineers. The American Society of Civil Engineers, which gathered these statistics, terms the current situation an "infrastructure crisis."

The Minneapolis bridge collapse provided dramatic evidence of this crisis. But it was by no means an isolated event. In March 2006, for example, an earthen dam in Kauai, Hawaii, gave way and let loose nearly 300 million gallons of water, killing seven people. In late 2005, a 120-ton concrete beam fell from a bridge in Pennsylvania onto Interstate 70. And of course, the levees in New Orleans were not only breached during Hurricane Katrina but structurally failed.

It is crucial that investments are made to upgrade the nation's crumbling infrastructure. In the meantime, however, more failures should be expected. The question now is whether we will be able to anticipate these failures in time to head off disaster. Sensor technology, if effectively implemented, would give reason for hope.

Good Enough for Government Work: A Model of Quality Management

I T IS A METAPHOR FOR OUR TIMES THAT THE AVERAGE AMERICAN HAS NEVER heard of Joseph Juran. Juran, who died February 28, 2008, at age 103, was a giant in the quality movement that revolutionized manufacturing, first in Japan and then in the United States and the rest of the industrialized world. Juran and those who followed him extended quality principles across the entire business sector and into other aspects of society. That his name is not better known in America today speaks volumes about this nation's failure to capitalize fully on his notions of teamwork, continuous improvement and quality-assurance management—notions that are more important than ever to the United States' position as an intellectual and practical world leader in innovation.

Born in a primitive East European village, Juran immigrated to the United States as a child, and at twenty-one was one of the first engineers to apply statistical methods to quality inspection in manufacturing. Ultimately, he became Western Electric's corporate head of industrial engineering and went on to re-engineer military logistics during World War II. On loan from Western Electric to the federal government, Juran led a multiagency team that redesigned the U.S. armed forces' shipping processes, reducing the amount of paperwork, significantly cutting costs, and aiding the war effort.

After the war, Juran became a full-time quality consultant and is credited with transforming the Japanese postwar economy. He popularized the Pareto Principle—the idea that 80 percent of potential improvements are due to 20 percent of operations—teaching that the most successful organizations optimize that vital 20 percent first.

Juran's key insight is that process matters. He stressed the importance of empowering individual workers, the reinforcing nature of teamwork and quality circles, and the importance of extending quality-management techniques to suppliers and customers. And he taught the importance of benchmarking to understand and meet the challenge of competitors. Japan's embrace of quality management placed it on the road to world manufacturing leadership, as documented in the book *The Machine That Changed the World*. Japan's recognition of Juran's contribution led to his receiving the Order of the Sacred Treasure, that nation's highest honor. This is why the Japanese were so amused when, in the 1970s, American companies wanted to learn Japanese management techniques. The Japanese believed they were practicing American management as taught by Juran and his colleague Dr. Edwards Deming.

Juran's book, *Total Quality Management*, is the bible on this topic. At age eighty-two, Juran was the star witness in the congressional hearings that led to the creation of the Malcolm Baldrige National Quality Award, which honors superior performance in organizations that function at the highest quality level.

Juran's Lessons for America Today

Management guru Peter Drucker stated in a 1996 PBS documentary that "whatever advances American manufacturing has made in the last thirty to forty years, we owe to Joe Juran." Indeed most large companies worldwide have embraced his ideas, including U.S. quality award-winning manufacturers Boeing Co., General Motors Corp.'s Cadillac unit, high-end textile and chemicals company Milliken & Co., and Texas Instruments Inc. Yet, U.S. manufacturing today accounts for only 15 percent of our gross domestic product. This is less than Japan, Germany, and other high-wage economies. Why? The reason is that services now dominate our economy, yet Americans often wonder where the service is in our service economy.

Quality principles apply in services as strongly as they do in manufacturing, but the problem is that much of our service and manufacturing sectors still cling to the rigid industrial efficiency production models of Frederick Taylor, the turn-of-the-last-century's most famous efficiency-management expert. The high-throughput Taylorist model that treats workers as automatons, quality as an after-the-fact consideration, and customers as uninformed and undemanding still appears to dominate.

Nor are we as a nation aware of how fast our competition is moving. Other countries are taking quality a step further by considering how their governments and educational institutions need to restructure to better accomplish their national goals. Taiwan's industrial parks now often bundle leading research universities and government agencies to provide research and policy expertise for integrated solutions. The result is a nation that has moved from an underdeveloped country famous for cheap goods to the world's largest manufacturer of all manner of computer peripherals—and increasingly the inventor, designer, and manufacturer of cutting-edge electronic technologies.

Or consider Finland, which has coupled its emphasis on quality in business with application of quality principles to its schools, with the goal of empowering its students and teachers. Despite de-emphasizing standardized tests, it has raised performance levels to the point where students are among the best in the world, both in standardized tests and adaptability in the work force upon graduation.

Countries across the globe are aggressively modernizing, and once again, as in the 1970s, the United States is not keeping up. That's why re-engineering government through quality management is so essential.

Public Policy Quality Management

Ten years ago, in the book *The Death of Common Sense*, legal scholar Philip K. Howard documented the rise of rules in American government. We were then, and still are, a society in which rules and procedures often inhibit creativity and problem solving. Much of government is driven by the same bureaucratic approach that Joe Juran spent a lifetime working to replace.

When people say they dislike government, they highlight the maze of rules and regulations, the lengthy and seemingly illogical processes, and the difficulty in getting governments to make decisions. Too often, government acts after the fact when something goes wrong—the analog of old-fashioned quality control—rather than working with its constituencies to avoid problems, an approach that lies at the heart of today's quality-assurance programs in the private sector.

Once something goes wrong, too often our government embraces an overbroad rule to prevent it from happening again. One shoe bomber, and thousands of Americans take off their shoes to comply with a rule that does not make us more secure and does not anticipate the next event.

Joe Juran and his colleagues have shown us the way out. It is now time for the government at all levels, and wherever feasible, to replace end-of-the-line regulation with active participation. A re-engineered regulatory agency should be able to deliver a higher level of public safety by working with companies and all other interested parties on setting commonly acceptable standards that guarantee a high level of public good in a cost-effective manner.

For instance, state auto inspections could be redesigned to increase public safety. Data that they routinely collect could be targeted to that 20 percent of components that cause 80 percent of the safety problems—the 80/20 rule again. Faulty components could be traced back to a manufacturing lot and allow auto companies and their suppliers to correct unusual patterns of wear. Currently, wide-ranging recalls happen only after significant and catastrophic failure. Why not solve the problems *in situ* and work for much smaller recalls well before failure occurs?

Critics will claim that the active participation of government agencies and industries in solving problems will result in what's known as regulatory capture, in which the industry being regulated commandeers the agency policing it. But that's what happens when regulations are rules-based; companies work hard to change the rules. In contrast, quality-assurance management, if properly implemented, replaces regulatory capture with cooperation and brings public officials, private industry, and consumers and workers together to achieve maximum efficiency, innovation, and speed in getting the product or service right the first time.

A government operating on its own priorities at a bureaucratic pace does not deliver timely solutions—just when U.S. international economic competitiveness demands timely action. To be effective in the twenty-first century, governments need to switch to a quality approach in conducting government business. This means going digital wherever possible, which in turn will necessitate setting privacy protection for the mining of that data, working with manufacturing and service industries on developing common language, and setting standards to make government/industry interactions as seamless as possible.

In short, it will require a government commitment to excellence in forcing out waste and maximizing efficiency in its functions, and in committing to what is best for private-sector entities and the common good. We need to realize that government services are part of the international competitiveness of private companies. Joe Juran championed international benchmarking for companies. We need to benchmark government delivery of services against the most efficient governments in the world. We need to get over assuming that everything in the United States is above average.

Finally, we must recognize our nation's inherent advantages and work to strengthen them as well. We have one of the world's longest traditions of democracy and of productively working together. The Internet has greatly increased the efficiency of democracy by rewarding open systems and by making distributive work and decision-making easier. Premium services from government are one of the offsets we can offer to low wages from other countries.

We have every reason to believe the United States can once again emerge as a world leader in productivity and quality of life if we focus on the vital issues where we can make the greatest improvements; if we err on the side of an open and free society; if we reorganize to empower our entire work force; and if we update Joe Juran's gift of quality through a commitment to focused, continuous improvement.

NURTURING THE NEW

America has for so long been recognized and even worshiped as the standard-bearer of scientific and engineering achievement that it can be difficult to adjust our self-image to accommodate a more shared, internationalist vision of progress. Yet everything about the way the global scientific enterprise is evolving points to the same truth: the world is becoming scientifically more integrated, and it will be ever more important in the years to come that we learn how to share and collaborate with our far-flung foreign colleagues in the sciences and other disciplines.

Whether through harmonized and updated methods of securing intellectual-property rights, new funding mechanisms that break down traditional silos among researchers in different cultures and specialties, or a willingness to rethink our immigration and educational systems to take advantage of—rather than recoil from—the many talented people from around the world who want to be part of the American scientific powerhouse, it is time to open our arms and our borders to a new way of fostering innovation here and abroad. As the following pieces show, there is plenty of inspiration to be had from the new ideas and approaches that are already bubbling up from beneath.

It Takes
a Village
to Issue
a Patent

T'S THE INCIDENT THAT STILL MAKES LIPS QUIVER ON EVEN THE MOST hardened Capitol Hill staffers. It was the winter of 2006, and it seemed likely that the BlackBerry would soon be silenced. The indispensable personal digital assistant was poised to fall victim to a multi-year lawsuit that pit its maker, Research in Motion Ltd., against a Virginia patent-holding firm. At issue: technology patents that cover the wireless sending of e-mails, many of which should never have been issued in the first place.

Research in Motion finally saved its precious BlackBerry by cutting a settlement check for $612 million, but the same sort of patent problems that threatened that device abound in the United States today. Indeed, software patents' fuzzy boundaries are widely reviled for stifling scientific innovation. But change is afoot.

Recently the U.S. Patent and Trademark Office (PTO) has been working with academics and the information-technology industry to test a plan that will be instantly familiar to anyone who has ever been to Slashdot, the online nerd community. Inspired by the decentralized Internet, the PTO's Peer-to-Patent project is raising hopes that one solution for tangled bureaucracy may well be putting the public at the levers of government.

Can it possibly work? Peer-to-Patent's trial run was admittedly small-scale. From June 2007 to April 2008, just forty applications were community vetted. But it worked well enough to earn it a second year of life and an expansion to include the related field of business methods, such as Amazon 1-Click, an online shortcut that makes buying books and other goods on the commerce site as simple as a single touch of the mouse. And as it grows, Peer-to-Patent raises the possibility that as the information economy gets ever more complex, the U.S. government won't insist upon making sense of it alone.

Overlapping patents issued by the PTO create what's known as a patent thicket, a deep and dark underbrush of conflicting claims that is enormously expensive to cut through. Research in Motion's out-of-court settlement check was cut even as the patents at issue were failing in the courts. Why has it been so difficult to avoid bad patents? The system has long been closed to outside help, and, internally at the PTO, it's in crisis. Seventy percent of patent examiners have considered leaving because of the unreasonable pace at which complex patents have to be considered. Just twenty hours or so are allocated for each application, according to a 2003 landmark report by the Federal Trade Commission on the economic cost of our patent woes. It's a pace set in the 1970s, when software was simpler and Windows was still a glimmer in Bill Gates's eye.

No matter how dedicated the agency's examiners might be, they are the single point of failure at the PTO, according to IBM Corp.'s associate general counsel, Manny Schecter, one of the Peer-to-Patent project's industry leaders. "We the people," says Schecter, "entrust the government in the form of the patent examiner to, in theory, know all the prior art that exists," referring to the record of past innovations that go to show whether an invention is novel. But with nearly a million patents backed up at the PTO's door, those examiners are being swamped by the deluge. And they've been suffering alone.

That is, until now. Peer-to-Patent grew out of a blog post written by New York Law School Professor Beth Noveck that was picked up by Wired.com, which in turn inspired a series of workshops. The PTO, ever eager to find a way out of the software-patent mess, jumped aboard. Put simply, the problem is one of information deficiency. Noveck frames it this way: "If you want to patent a battery-powered golf-club-slash-weed-wacker" (a genuine product marketed as the Big Daddy Driver: U.S. patent no. 6,988,954), "then you can find the state-of-the-art in golf technology just by looking at the patent database." By contrast, software—which has only really been patented since the late 1980s—lacks many of the terms of art of other scientific fields and is often undocumented and unpatented to begin with.

Lacking a robust patent record to draw upon, software-patent reviewers were lost. The Peer-to-Patent solution builds on ten years of lessons learned from the interactive Internet. It says, in essence, "Let's crowd-source it." Applicants seeking a software patent opt in via a simple form, after which their proposed patent is posted to peertopatent .org. The program is structured to attract citizen-experts, such as scientists, PhDs, and programmers, all of whom can vet the application's claims by submitting prior art that is either supportive or debunking.

Reviewers use Digg-like tools—which allow online communities to boost news stories through individual ratings—to vote on the best prior art, which goes to a PTO examiner for final determination. Along the way, these Peer-to-Patent reviewers tag the application's claims with keywords. The hope is to use the social Web's ideas about folksonomy, or the collaborative creation of taxonomies, to evolve an accepted lexicon for the field. It's really, says Noveck, "about opening the conversation about science."

If all goes well, a patent examiner uses his or her limited time judging the merit of the application rather than hunting for prior art. And the successful patentee gets the confidence that comes with having withstood peer review.

That anyone can participate, either by name or anonymously, naturally raises fears that it's an excellent chance to sink a rival innovator. But inventors, so far, are for the most part unworried. Nanomaterials researcher Blaise Mouttet, a Peer-to-Patent applicant, finds assurance in the idea that a patent yea or nay is "a fact-based determination." He wants, he says, "the best possible prior art so that I get the best possible patent." Indeed, bad patents have a way of revealing their flaws eventually, and it's better for a weak patent to die an early death.

Some, of course, are quick to warn that Peer-to-Patent is limited in what it can accomplish. Boston University School of Law's Michael Meurer, co-author of the influential jeremiad *Patent Failure: How Judges, Bureaucrats, and Lawyers Put Innovators at Risk*, is more or less a fan of the project but still worries that there's only so much that better prior art can solve when it comes to software patents. The seemingly intractable problem, says Meurer, is that when it comes to the software field, patents are "intrinsically vaguer and more problematic" than in other fields.

Meurer's concern points to the modern controversy over software patents in the United States. Does software—or, more accurately, the techniques and features that are the building blocks of software—really have any business being patented in the first place? In the United States, the full flowering of a software developer's expression is, of course, the program. And programs are already protected under the Copyright Clause of the Constitution: "To promote the Progress of Science and useful Arts, by securing for limited Times to Authors and Inventors the exclusive Right to their respective Writings and Discoveries."

The problem, says Meurer, is that patents on software and business processes "cover what the invention can do." And "can do," he says, is a shaky foundation on which to build a system designed to encourage innovation. To give one familiar example, Adobe's ubiquitous Acrobat program is already covered by copyright. So is it really necessary to grant monopoly patent rights as well to the more than 65 different fuzzy-boundaried bits that make up the program?

It's a debate that is currently raging in Europe, where the idea that software processes are a poor fit for patent law has maintained a slim edge over the contrary view.

Richard Stallman is a famed software guru who founded the Free Software Foundation to fight just such fights. Stallman is a harsh critic of Peer-to-Patent, condemning it as a tool that only empowers wrongdoing. "This system will help patent applicants gerrymander their patents so that they resist prior art, while still being useful weapons to attack" software programmers and developers.

The goal, Stallman argues, should not be to fix up the software-patenting process, but to burn it to the ground. "The idea of 'Peer-to-Patent' presumes that patents are good and that the goal is to implement the system 'better,'" he explains. "In some fields that may be true, but not in the software field. Software patents must be eliminated entirely."

It's an all-or-nothing argument familiar to Mark Webbink, the former intellectual-property officer at open-source software powerhouse Red Hat Inc. Webbink now oversees Peer-to-Patent's expansion at New York Law School. He shakes off the criticism that Peer-to-Patent is wasting time bettering a system that should be torn down. "I've made it pretty clear in my career that I'm not enamored with the patent process," he says, in a characteristic bit of understatement. While at Red Hat, Webbink mocked the patent-zealous in his industry, including Microsoft, whose pursuit of patents relating to the management of extra space in documents has gone so far as to include separate filings for methods for adding white space and for deleting it.

But the United States today, explains Webbink, just isn't amenable to black or white solutions. "Minimal reform legislation is tough," he says. "Major reform legislation is damn near impossible." The powers that be, from Congress to the courts, are thoroughly enamored with the idea that patenting software is necessary to keep America's high-tech innovation chugging along. "You can hold your breath until software patents go away," says Webbink, "but you're going to expire before they do."

John Doll, who served as U.S. Commissioner for Patents under George W. Bush, has argued that when patent examiners get good prior art in front of them, they make the right decisions. "If the public has criticisms" about the software-patenting process, he has said, "this is the opportunity for them to step up." Log in to Peer to Patent, and be part of the solution.

And why stop there? After all, while Peer-to-Patent is, to be sure, an exercise aimed at fixing the patent process, it's also an experiment in using the tools of the social Web to create models for participatory government. "We could substitute almost any area of policy that depends on good information to make a decision," says Noveck, whether it's a patent examiner facing an enormous stack of software-patent applications or a Capitol Hill staffer staring at a 300-page energy-reform bill. Could Peer-to-Patent be pointing the way toward how we can engage the public in government?

When we talk about citizen participation in government, we're often talking about tapping into people's feelings and judgments, says

Noveck. Peer-to-Patent works by focusing on fact-based expertise. "The idea of a free-for-all, of putting something up on a wiki," Noveck says, "is terrible." But civic engagement online starts to become useful, she says, when you design a system with structured roles for people with expertise and group checks on individual contributors.

At its core, Peer-to-Patent starts with an admission that one particular bureaucracy needs a helping hand. "It has to be okay for a patent examiner to say, 'I don't know,'" says Noveck. "And that needs to be true across government." The first step toward a government that can cope with the complexities of the modern world might well be acceptance of the fact that it can't do it alone.

Riding
the
Sci-Tech
Express

A S IT ENTERS THE TWENTY-FIRST CENTURY, AMERICA FACES SIGNIFICANT new challenges, including threats to U.S. economic leadership, growing entitlement and health-care spending, and the many costly consequences of climate change. Meeting these challenges will require considerable political will. But equally important, it will require a renewed commitment to science and technology.

America has enjoyed a remarkable, decades-long increase in its standard of living and economic competitiveness, and scientific and technological innovation has been at the heart of that trend. By some estimates, according to the prestigious National Academies, about half of all U.S. economic growth since World War II has been the result of technological innovation. Even more impressive, virtually all of the acceleration in productivity growth since 1995 has been due to the information-technology revolution. Technological innovation will become even more important as we struggle to compete successfully with other nations and sustain the growth rates needed to support our growing numbers of retirees.

So how do we bring science and technology to bear on these and other challenges? We can start by realizing that science and technology should not be the province of a couple of congressional committees with the term "science" in their title. Whether the problem is health care, climate change, renewable energy, national security, education, or the economy, science and technology will be central to success. And that will require across-the-board commitments to boost investment, enhance collaboration, and cultivate talent.

More Investment

In the past, when the United States faced challenges that could be addressed by science and technology, Americans acted with resolve. In the decade following the launch of Sputnik, for example, federal support for research and development doubled (in constant dollars). In contrast, over the past decade—and in the face of challenges arguably more profound than the launch of Sputnik—federal support for research and development increased just 25 percent, and as a share of gross domestic product it actually went down.

In fact, the United States is one of the few nations in which total (that is, combined public and private) investments in R&D as a share of GDP fell from 1992 to 2005. To restore federal support for research as a share of GDP to the level it was during President Bill Clinton's first year in office, we would have to increase federal R&D spending by 50 percent, or more than $37 billion.

Unfortunately, the modest increases Congress is now considering, while a step in the right direction, are woefully short of what is needed. It's time to make a national commitment to increase federal support for research by $3 billion per year for the next decade. This would be equivalent (again, in constant dollars) to the increase in federal support for R&D in the decade following Sputnik.

One way to pay for at least a portion of this increase would be to impose a carbon tax, perhaps in conjunction with a carbon cap-and-trade system, and devote a significant share of that revenue toward clean-energy R&D, one of the best investments in this nation's future that we can make today. But it is not sufficient to simply increase public support for R & D. We also need to encourage businesses to do more R&D, especially at home. Consider that from 1998 to 2003, investment in R&D by U.S. companies increased twice as fast overseas as it did in the United States: 52 percent versus 26 percent. The attraction is obvious. Not only are wages of researchers in places such as India and China less than one-fifth as high as they are here, but many foreign nations now provide significant R & D incentives for U.S. companies to conduct their research abroad.

As a decade's worth of economic research has shown, increasing the corporate tax credit for technology R&D would spur companies to conduct more research in the United States. When Clinton took office, the United States enjoyed the most generous R&D tax credit in the world. Today, we are sixteenth, behind countries as different by levels of development as Mexico, Canada, Japan, and France. It's time to bring the R&D tax credit into the twenty-first century by doubling the regular credit and the Alternative Simplified Credit, which can help companies that are not eligible for the standard credit. Doing so would make an important statement that the United States is serious about keeping and growing research-based economic activities.

More Collaboration

If the United States is going to meet today's looming economic challenges it will have to move beyond its traditional focus on funding basic science and start promoting innovation partnerships between the government, the private sector, and universities.

Other nations have come to that conclusion. In recent years, Finland, France, Iceland, Ireland, Japan, the Netherlands, New Zealand, Norway, South Korea, Sweden, Switzerland, and the United Kingdom, among many others, have either established or significantly expanded various innovation-promotion agencies. It is time for the United States to do the same, and to create and fund a National Innovation Foundation, as a number of progressive institutions have recommended. The foundation would neither run a centrally directed industrial policy nor dole out corporate welfare. Rather, it would work cooperatively with individual businesses, business-university consortia, and state governments to foster the kind of collaboration and innovation that would benefit the nation but would otherwise be unlikely to occur.

In addition, Congress should allow businesses to take a flat credit of 40 percent for collaborative research conducted at universities, federal laboratories, and research consortia. This would spur more research partnerships between companies and American universities and federal laboratories.

More Talent

We now lag behind much of the world in the share of college graduates majoring in science and technology. The United States ranks just twenty-ninth out of 109 countries in the percentage of twenty-four-year-olds with a math or science degree.If we are going to make a significant push to expand science and technology to solve these new challenges, we will need a significantly larger supply of home-grown talent in science, technology, engineering, and math (STEM), as well as policies that make the United States a more attractive destination for foreign scientists and innovators.

Congress took steps toward strengthening domestic science education by recently passing the America COMPETES Act, which among other things offers support for more specialty math-and-science high schools. But this legislation needs to be expanded and fully funded.

Even with these efforts, however, it's important to realize that, at least for the short term, we won't be able to rely on domestic supply alone. As a result, Congress should expand and reform the H-1B visa program for skilled technology workers, to attract more talent and to ensure that employers are paying prevailing wages. Finally, immigration policy should be adjusted to make it easier for foreign students studying in STEM fields to attend school here and to gain a path to citizenship once they obtain their graduate degrees.

Despite their great need, enacting these and other pro-science and technology policies will not be easy. In fact, disappointingly, there is some resistance to these policies across the political spectrum. Some fiscal conservatives see the role of government in science and technology as best limited to basic research, wrongly presuming that industry will reliably step forward from there, even though profits are in many cases uncertain. Moreover, for many of these conservatives, an unswerving commitment to tax cuts for individuals undermines the government's ability to fund work in science and technology. Then there are the social conservatives who oppose federal support for research they personally disagree with, such as studies on stem cells and therapeutic cloning, or who oppose public funding of the teaching of evolution.

Conversely, many social liberals look at science and technology with suspicion, and are particularly skeptical of many biotechnology, nanotechnology, and information-technology innovations. Underpinning their opposition is a naïve and elitist belief that we can somehow hold back the forces of progress and even go back to a "simpler" time. Environment activist Bill McKibben sums up this view when he states, "'More' is no longer synonymous with 'better'—indeed, for many of us, they have become almost opposites."

Yet try telling that to average American workers struggling to make ends meet and afford a mortgage, health care, and college education for their kids—and perhaps even daring to wish for a nice vacation. Without the new products and services from scientific innovation and higher productivity—innovation, remember, is the source of virtually all productivity growth—the American standard of living and quality of life are bound to stagnate, and the economy will become less and less competitive vis-à-vis other nations. It is terribly important that legitimate concern for fiscal discipline not preclude consideration of significant new investments in science and technology, which history has shown are sure to pay off with large and in many cases unexpected dividends.

In spite of the challenges, we can find our way forward to create a political consensus that is not just supportive of science and technology but downright passionate. Doing anything less will mean that we give up on what has made America unique: our faith in the future and our belief that technological creativity and innovation will produce a better tomorrow.

Securing
Our
Scientific
Future

I T HAS BEEN SO LONG SINCE THE UNITED STATES HAD TO LOOK UP TO ANY country in science that we Americans have come to regard science leadership as a birthright. When children in other countries score better on science tests than American youngsters, or our production of PhDs and engineers or share of patent applications declines relative to other countries, we act as if the United States is slipping rather than other countries advancing, and we see a crisis emerging.

Perhaps the field we have truly fallen behind in is history. We forget that in the early twentieth century, German was the lingua franca of science. Germany was where young scientists went to study, and where top scientists presented, and often did, their cutting-edge work.

Far from being natural or inevitable, U.S. science leadership today is an offshoot of this country's pre-eminence after the defeat of Germany in World War II. It reflects the waves of innovation that have been spawned by America's vibrant capitalist economy, spurred in particular by WWII—and subsequent Cold War—impelled needs to develop our science and engineering capacity. It uniquely benefited from immigration, especially from Europe in the Nazi era and the immediate postwar period. And it could not have happened without the wealth and vision that allowed the United States not only to generously subsidize basic science but also to establish an educational system that was broad-based at the bottom and unparalleled in availability and quality at the top.

If these advantages were not enough, the competition for science leadership was weak thanks to the devastation that Europe suffered in two world wars and the slow rebuilding of European economies in the postwar era.

But while today's U.S. science leadership is neither natural nor inevitable, the loss of that leadership may be. Countries much larger than the United States, most notably India and China, are experiencing economic growth that outstrips ours, and as they grow in wealth they are rapidly improving their educational systems and basic-science infrastructures. Moreover, as globalization leads U.S. companies to move research and production capacity abroad, market demand for trained scientists and engineers is increasing elsewhere while being dampened here.

Even if the United States retains a per capita education and investment advantage over India and China, population differences alone mean that the number of trained scientists and engineers in these countries will soon dwarf the number in America, with differences in the quantity and quality of science innovation likely to follow. Added to the Asian challenge is a Europe that can no longer be seen as a set of discrete countries when it comes to science. Rather, cross-border research teams are being encouraged, and European Union—wide funding mechanisms are being established.

In short, several decades from now we may find that we are not the world's no. 1 country when it comes to science, however measured, but perhaps no. 4 behind China, India, and the EU. We may also find that being in fourth place is not altogether bad. When children in China are vaccinated against polio, they are not worse off because the vaccine was invented in the United States. When an Indian inventor draws on two decades of U.S. government—funded research to achieve a technological breakthrough, her accomplishment will not be lessened because of its early dependency on U.S. basic research.

U.S. science investments have had substantial spillover effects, improving the quality of life in other countries and enabling scientific, technological, and medical accomplishments that have benefited people abroad. Similarly, as other countries improve their science, the progress of American science and the lives of our people will increasingly benefit from educational and infrastructure investments made in those countries and from research supported by currencies other than the dollar.

Acknowledging the possibility of America's loss of science leadership and seeing a bright side does not, however, mean we should regard that future as an unalloyed blessing and passively allow American science to slip. There would be substantial costs were U.S. science capacity to sink absolutely, and real costs even if slippage is only relative. Scientific advances create intellectual property, and wealth creation through intellectual property has become an increasingly important part of the U.S. and world economies. What's more, the world remains a dangerous place, and it may become more so should countries like China develop expansionist ambitions. Science for security must remain a high national priority, and although we may not be able to keep other nations from catching up, we should not be complacent about the possibility of their surpassing us, and we certainly do not want to find ourselves surprised by their achievements.

In devising policies to maximize the strength of U.S. science, our nation has two unique resources it must not squander. The first is English. Thanks to the pre-eminence of U.S. science for more than half a century, English is second only to mathematics as the universal language of science. Scientists around the world speak and write English. This gives American scientists a leg up in communicating with scientists across national boundaries and makes many of the most important writings of foreign scientists easily and immediately accessible to Americans.

Additionally, American students are not dissuaded from pursuing science careers, nor do they have their science studies delayed because of the need to master a foreign language. Short of eliminating federal science funding, probably nothing would harm American science as much as a need to read Chinese to keep up with the latest science developments.

One goal of our national science policy should be to maintain English as the global language of science. This might entail subsidies or other incentives to promote the publication of English-language online science journals, aid to enable the acquisition of English-language science materials (including print journals) by universities and libraries abroad, and programs to train foreign scientists in English, either in their own countries, online, or by bringing them to the United States or Britain for science internships or language instruction.

The high subscription price of leading English-language science journals is a particular threat because it means that for financial rather than science reasons market forces are likely to promote a proliferation of lower-priced, foreign-based journals in languages other than English. These journals, started for reasons of cost, may become science journals of record in their home countries, meaning that cutting-edge overseas research may become less easily or immediately available here. The short-run solution may be U.S. subscription subsidies for foreign scholars and institutions, but the only viable long-term solution is to bring costs down, most likely by electronic distribution that through competition reins in the profit-oriented publishers who now mediate between the creation and distribution of science knowledge.

The United States' second great advantage is our system of higher education. We are still the pre-eminent nation when it comes to science training, and we benefit from this in many ways. Foreigners who come to study here learn English, and they build relationships with U.S. scientists that endure after they return home, if they return home. Study here can also lead to an appreciation for the United States and its values, including especially the values of democracy and free inquiry. Perhaps most beneficial of all are the foreign-born scientists who stay to take jobs here or who return periodically to work collaboratively with U.S. scientists. They add to our science work force and scientific productivity and go a long way toward making up for inadequacies in the production of U.S.-born scientists.

Ironically, the threat to U.S. science dominance is in part due to our willingness to educate the world. Some of the foreign scientists trained here have returned home to become leading researchers or educators in countries such as India and China, while others have returned to Western Europe and reinvigorated graduate science education there. Thus, our leadership in science education, although not as vulnerable as our overall science leadership, is also ripe for challenge.

Rather than rise to the challenge, however, we have aided the challengers. Short-term political and security concerns relating to the "war on terror" have trumped longer-term interests in science strength, along with longer-term wealth and security. Responding viscerally to the attacks of 9/11, we made entering this country more difficult for foreigners whatever the reason. One result was that students who had planned on doing their advanced science studies in the United States went instead to Europe, Australia, Japan, or Canada. Or they pursued advanced degrees in their home countries.

More recently, the Iraq war and ungenerous attitudes toward immigrants have made the United States less attractive to educated foreigners. Difficulties in entering the United States have also affected the location of and attendance at scientific conferences, as well as the ability of universities and companies to employ foreign researchers. Although the U.S. government has become sensitive to the harms that some of its post-9/11 policies caused and has tried to ameliorate problems, it could be doing much more—including proactively encouraging more foreign students to study science here and making it easier for them to work here when their studies are concluded.

The downside of replenishing our science work force with the foreign born is that it diminishes pressure on industry and government to stimulate domestic science training. Yet few dispute that improving domestic education must remain a high priority, especially as opportunities for science workers abroad grow sufficiently attractive as to not only lure foreign-born U.S. science workers back to their home countries, but also to entice native-born American scientists to work abroad. And that must include a real focus on populations that have until now not been sufficiently courted for futures in science. We cannot afford to leave undeveloped the talents of minorities and the poor by failing to provide the nutrition, health care, preschool training, and later education that would allow these youth to realize their potential. It is no longer just a matter of personal achievement or social justice, but rather one of economic and national security.

If we educate our youth in the sciences, if our science work force is steadily recharged with fresh thinkers from home and abroad, and if we facilitate the international exchange of scientific knowledge, then we will not have to worry about whether other countries are doing as well as or better than we are. We will be strong as a nation, and we will remain a fountain of innovation into the future.

ADVANCING THE INNOVATION ECONOMY

By many measures, and particularly as a percentage of gross domestic product, America's investment in research and development has decreased while that of our competitors continues to grow. During an economic downturn it's hard to focus on the long run, and in fact, technological innovation does not in itself pull an economy out of an acute tailspin. That takes time, and some global cooperation.

All the more reason for government to exert leadership by removing obstacles, providing incentives, and serving up some old-fashioned reminders about our obligations to future generations. The research horizon is usually not a mere two-year or five-year commitment, but must normally be measured in decades. Yet for any twenty-first century country that aims to continue to be a global leader, there is no choice but to invest in innovation. How to do that in the smartest way possible is the subject of this section.

Rx:
Innovation
Infusion

EVERY TWO YEARS THE NATIONAL SCIENCE BOARD, WHICH CONGRESS established in 1950 to provide advice to the president and Congress, releases a report called *Science and Engineering Indicators*. It is not exactly a best seller, but it is a consistently treasured resource for policymakers because of its in-depth assessments of the state of science and engineering research and education in the United States.

In 2008, however, the board, which in addition to advising America's political leaders serves as the governing board of the National Science Foundation, felt compelled to release a companion report to highlight the policy implications of its findings in that year's *Indicators* report. Too much was at stake to simply release the latest figures without additional contextual information.

Keeping in character, the title of that report from the board did not invite casual readers. But *Research and Development: Essential Foundation for U.S. Competitiveness in a Global Economy* is an important and prescient volume for anyone who cares about the economic future of the United States. It concludes that the current decline in support for basic research by both industry and government stands to severely harm U.S. competitiveness in international markets and undercut highly skilled and manufacturing jobs at home.

Among other worrisome findings, the board noted that the U.S. high-tech trade deficit jumped to $132 billion in 2005, the last year complete data were available, from $32 billion in 2000. At the same time, federal support for academic research and development began falling in 2005 for the first time in a quarter-century. And private-industry support for basic research has proven to be at best stagnant, and possibly declining. Basic research published in peer-reviewed journals by industrial researchers declined 30 percent in the last decade. And the number of physics publications by industrial researchers dropped to only 300 in 2005, from nearly 1,000 in 1988.

Another study commissioned by the National Science Foundation concluded that if current trends continue, "China will soon pass the United States in the critical ability to develop basic science and technology, turn those developments into products and services—and then market them to the world."

These reports strengthen the case for bold, decisive action to restore America's scientific and technological leadership. In *A National Innovation Agenda*, a report published in 2008, the Center for American Progress offered a detailed set of policy recommendations for doing just that. As part of the center's economic plan for the next administration, it called for:

- Doubling the research budgets of key science agencies such as the National Institutes of Health, the National Institute of Standards and Technology, and the Department of Energy's Office of Science, and providing even larger increases (10 percent per year) for the National Science Foundation and the Defense Department.
- Maximizing the effect of this investment by increasing support for university-industry collaborations and high-risk, high-return research. This would help replace the void left by the decline of industrial basic research documented by the National Science Board.
- Harnessing science and technology to address some of the "grand challenges" of the twenty-first century, such as the transition to a low-carbon economy that will reduce our emissions of greenhouse gases while creating millions of "green collar jobs," or the development of new learning technologies that are as effective as a personal tutor and compelling as the best video game.
- Spurring private-sector investment in innovation by making the research and experimentation tax credit permanent and

providing tax incentives for investment in next-generation broadband networks.

- Ensuring that America's work force has world-class skills in science, technology, engineering, and mathematics (STEM). This will require upgrading the STEM skills of the existing work force, improving K-12 math and science education, encouraging more students to receive undergraduate and graduate degrees in STEM fields, and creating a fast-track, employment-based visa for foreign students who receive advanced technical degrees from U.S. universities.

The latest findings from the National Science Board echo those from a long series of reports that have documented the erosion of America's scientific and technological leadership in the past decade. Unfortunately, although President George Bush and Congress pledged in 2007 to address the problem with the American Competitive Initiative and the America COMPETES legislation, both branches failed to keep those promises.

The National Science Foundation's research budget received a paltry 1 percent increase that year, the budget of the National Institutes of Health is down more than 10 percent in real terms from 2004, and the research and experimentation tax credit was allowed to expire in December 2007 for the thirteenth time. At the same time, Congress slashed funding for the Department of Energy's high-energy physics program, forcing gold-standard research institutions such as the Fermilab, the Argonne National Laboratory, and the Stanford Linear Accelerator Center to slow or stop important projects and fire scientists.

This failure to act is unacceptable. It is critical that our new president make science, technology, and innovation a top priority. The future economic prosperity of our nation and our people depends upon it.

Don't Cry for Me, EVITA

HERE IS ONE SAFE ASSUMPTION ABOUT THE FUTURE OF U.S. ECONOMIC prosperity: without our nation's robust venture capital-backed, entrepreneur-driven, tech-flavored industries and services, which are today still largely unique to the United States, our country is destined to become an also-ran, its global influence, power, and leadership fated to diminish over the course of the twenty-first century.

That truth is an important one to acknowledge because our clear lead in what some have begun to call the EVITA economy—entrepreneurial, venture-backed, information-dependent, technology-flavored activity—now faces rising challenges from global competitors. Monopolies never last forever, of course, and competition is a natural, inevitable, and even healthy phenomenon. Indeed, the good news is that an increasingly global, integrated, innovation-led economy is certain to bring prosperity to both our country and others given the tsunami of scientific and technological advances underway.

Unfortunately, serious challenges—many of them of our own making—today threaten the very keystones of our nation's entrepreneurial economy: startup technology companies. The reasons are manifold but largely stem from insufficient allocation of financial and human capital and misguided policymaking. Specifically, what this country lacks today is an economic policy designed to grow innovative new tech companies intelligently and productively, yielding benefits that are socially positive, widely distributed, and fairly allocated.

The United States is unusual in that many of its strategically vital corporations are young and competitive, contributing enormously to our national prosperity. In 1990, Microsoft Corp., Dell Computer Co., and Cisco Systems Inc. had combined sales of $2 billion. By 2000, their combined sales hit $80 billion before leveling out at around $90 billion in this decade as they gained blue-chip status. Such growth exemplifies why companies backed by venture capital generate twice the sales, pay three times the federal tax, and invest far more heavily in research and development than their traditionally financed counterparts.

Specifically, companies backed by venture capital generate $643 in sales for every $1,000 in assets, compared with only $391 in sales for traditional companies. Venture-backed firms also spend about $44 per $1,000 in assets, compared with $15 by others. And recent statistics show that approximately 11 of every 100 working adults in the United States were engaged in entrepreneurial activity, either starting a business or playing a lead role in one less than three-and-a-half-years old, a rate higher than any in Europe and roughly twice that of Germany or the United Kingdom.

"Entrepreneurship is what enables American-style capitalism to be generative and self-renewing," observes Carl Schramm, head of the primary U.S. research institute on venture capital, the Kauffman Foundation. But disappointingly, Schramm adds, "The system that generates and supports entrepreneurship in the United States is surprisingly unappreciated."

The statistics deserve a wider audience. For three decades, venture capital-backed startup companies have been the job-creating engine of the U.S. economy. According to a study by the consulting firm Global Insight, released by the National Venture Capital Association, startups backed by venture capital since 1970 today employ 10 million Americans and in 2005 generated sales of $2.1 trillion. These companies employ more than 9 percent of the U.S. private-sector work force and account for an astounding 16.6 percent of gross domestic product (GDP).

This is amazing when you consider that the $23 billion invested by venture capitalists in 2005 represented only 0.2 percent of GDP. Talk about a bang for your buck. These companies—from widely recognized giants such as Apple Computer Co., Intel Corp., Cisco, Amgen Inc., FedEx Corp., and Google Inc. to lesser-known but up-and-coming mobile technology and lifesaving drug and device companies—have generated far higher than average wage growth, have accounted for a significant and growing proportion of U.S. civilian research and development, and have spawned some of the most innovative products, services, and business models of our era.

But we ain't seen nothing yet. Giant technological leaps are in their infancy. Fossil fuels will be, in fact, fossils. Management-information systems will revolutionize health care as computer-driven artificial intelligence facilitates diagnoses. Across our economy, one "pipe" will carry interactive video, audio, and the Internet into every home. Global positioning system-driven satellite systems will drive—literally!—our terrestrial, air, and sea vehicles and vessels. And these examples do not even start to include the vast changes afoot at the intersection of technology and biology.

According to world-class physicist Freeman Dyson:

> Two facts about the coming century are agreed on by almost everyone. Biology is now bigger than physics, as measured by the size of budgets, by the size of the work force, or by the output of major discoveries; and biology is likely to remain the biggest part of science through the twenty-first century. Biology is also more important than physics, as measured by its economic consequences, by its ethical implications, or by its effects on human welfare.

These facts raise an interesting question. Will the domestication of high technology, which we have seen marching from triumph to triumph with the advent of personal computers, GPS receivers, and digital cameras, soon extend from physical technology to biotechnology? The answer to this question is starting to look like "Yes." In fact, many scientists now predict that the domestication of biotechnology will dominate our lives during the next fifty years at least as much as the domestication of computers has dominated our lives during the previous fifty. For starters, some have gone so far as to believe that if all we do is bring to fruition some of the aging-related biotech innovations currently under development, then an average and active life span of 100 years is sure to follow.

Yet serious impediments exist in our capital markets that could inhibit venture capital-backed startups from commercializing these existing and future technologies. There are too many notional "public" companies whose shares are too small to trade unless promoted by unscrupulous crooks. There are too few opportunities for young high-tech companies to go public through an initial public offering without sharing a stock market listing with these crooks. There are not enough equity-flavored forms of compensation—options—for gifted managers. And there are not enough sensible tax incentives.

The first challenge, then, is for public policymakers to examine federal regulations so that our economic and financial policies foster entrepreneurialism, not only to preserve the EVITA economy in its present form—by attacking bureaucratic micromanagement that violates the principle of *primum non nocere* ("first do no harm")—but also to unshackle capital-raising opportunities for young tech startups willing to put in the necessary elbow grease.

Policymaking opportunities to boost innovative companies would facilitate the flow of more private capital into early-stage startups. And good policies would reform the irrational blockage in the pipeline between university research labs and venture investors, an aberration that has minimized the ability of universities and medical schools to commercialize technology through spinouts in which the lab owns a meaningful equity interest.

It's also important to enable investors below the level of multi-millionaire angels to diversify a prudent portion of their investment portfolios into well-managed venture funds through 401(k) and IRA pension plans by tweaking the Business Development Company amendments to the Investment Company Act. And we need to spread venture capital-driven entrepreneurialism beyond the East and West coasts into communities across the United States by promoting regional innovation centers of excellence.

There is growing consensus in the country and in Congress that scientific and technological innovation can be the twin engines of an economic recovery in this country. It is time to make a concerted effort to create more venture capital-backed opportunities for the ambitious and creative entrepreneurs across our country.

Old States Can Learn New Traits

U NTIL THE MID-1800S, PAPER WAS MADE MOSTLY OUT OF RAGS. THE paper industry lived and died on the availability of discarded clothes and other cloth materials that could be beaten into a pulp and poured into molds to make paper.

Papermaking began in Maine in the 1730s, as the state's rivers provided power and clean water for the process. The industry in general was successful, and Maine papermakers, like those in many other states with clean rivers, made a good living.

But the state's paper boom didn't really begin until a rag shortage in the 1850s, along with increasing demand for paper, forced European and American inventors to search for new resources to make pulp. They quickly discovered that wood was an efficient resource for paper, and wood-based paper mills began to spring up, primarily in the Northeast.

Developers saw an opportunity in Maine, the nation's most densely forested state. The first wood pulp in Maine was produced in the basement of a sawmill in 1868, where the workers made one ton of pulp per day. By 1875 the S. D. Warren mill in Westbrook was blending wood fibers with rag pulp, and five years later it was the largest paper mill in the world.

It was one of those rare confluences of a need, an idea, an abundance of resources, and the money to make it happen all meeting in the same place at the same time—the perfect formula for innovation to flourish. That confluence allowed the paper industry to serve as the primary driver of Maine's economy for the next 150 years.

But times change. As globalization makes it more and more diffi-cult for Maine to compete as a paper producer, the state is looking to preserve the industry of the past while moving to a new economic driv-er for the future. Other traditional Maine industries—like fishing and farming—are struggling, and rely on innovation to remain competitive and to grow into the twenty-first century.

Last year, the Maine legislature passed a historic bond package to send out for ratification by voters, who approved $50 million in com-petitive grants for research and development to be issued through the Maine Technology Institute.

The institute will award a mix of large and small grants to entrenched and fledgling research firms that make the best case that their invest-ment will provide the biggest return. With a total of $50 million avail-able to the best and the brightest—and at least $50 million in matching funds expected for winning grants—there will be no shortage of fresh ideas and innovative pitches. Some of these ideas will venture boldly into uncharted territory; others will improve old technology to pre-serve and improve traditional state industries like pulpmaking, fishing, and farming.

Historically, state investments in research and development have paid off. The Jackson Laboratory, based on Mount Desert Island in Penobscot Bay, is now the world's largest mammalian genetics research institute. Jackson Labs has been a regular beneficiary of state invest-ment, and it has developed into one of the world's premier biomedical research facilities. Targeted investments in boat building and marine research and development have established the state as an interna-tional leader in those industries as well.

To bring the paper story full circle, some new R&D money is likely to fund groundbreaking research happening at the University of Maine on turning wood product waste cellulose into ethanol. Scientists have discovered that wood ethanol is far cheaper to produce than ethanol made from corn. This research will likely add value to the bottom line of Maine's current paper industry and simultaneously reduce the state's carbon footprint.

In the late 1960s, mothers of students at a prep school in Seattle used proceeds from the school's rummage sale to buy an ASR-33 tele-type terminal and a block of computer time on a General Electric computer. For such a small investment, the computer was an enormous hit with the junior high school students, and access time to the GE computer was in high demand—especially for a group of thirteen-year-olds who saw a world open before their eyes.

Among them was one student who discovered an almost supernatural knack for deciphering and writing codes. He founded his first software company at age fourteen, developed payroll-processing and traffic-counting software through his teens, and at age twenty founded Microsoft—which went on to become the biggest software company in the world.

Bill Gates knew that he excelled at science and math from a young age, but he didn't know the potential behind it until a small investment made at the right time with rummage-sale proceeds allowed him to realize it. It's one of innovation's most famous success stories, and it's based on the same basic concept of a need, an idea, and a resource all meeting at the right time.

In the same spirit, the competitive grants awarded by the Maine Technology Institute will go to the people with the best ideas who can promise the best return on an investment. The institute is tasked with spotting potential—a challenging, exciting, and essential charge. Other states would do well to follow its example, as a number have begun to do in recent years, in their own ways.

Somewhere in Maine, and indeed in each U.S. state, is the next paper industry, the next auto industry, the next fishing and lumber and mining industries—future engines of growth that could drive local and statewide economies for centuries. Somewhere in Maine, and indeed in every U.S. state, are fresh young versions of Bill Gates, each just a small investment away from discovering an ability to change the world.

Diversity as a Driver of Innovation

WHETHER PUT FORTH BY VANNEVAR BUSH, ONE OF THE GODFATHERS of the military-industrial complex in the 1950s, or Richard C. Atkinson, the head of the National Science Foundation in the 1970s, or Francis Collins, who oversaw the massive and revelatory Human Genome Project in the late 1990s, the case for government funding of basic science has remained largely the same after more than a half century: investment in basic research produces the knowledge that drives innovation and, in turn, human progress.

This faith in public funding of basic research rests in large part on the proven shortcomings of profit-driven science. The demands of the marketplace obviously create incentives to explore many scientific pursuits, most often practical problems such as how to build faster computer chips, safer allergy medications, and more fuel-efficient cars. One fact should be obvious, though: businesses won't set out to make the world a better place unless they are likely to make money at it.

That leaves many fundamental questions related to the causes of disease, the forces that create an affluent society, and the maintenance of the Earth's ecosystems in need of funding from governments. Finding solutions to existing problems is reason enough to support science research, yet government investment in basic science also encourages unguided exploration that can result in solutions in search of problems, the laser being a prime example. As odd as it may sound, science often finds answers to problems we didn't even know we had.

Unfortunately, although the basic argument for government research-and-development funding has not changed in decades, neither has the means by which the government goes about that funding. Indeed, the current science-funding model, with its general insistence on a proven track record and a high assurance of success, is clearly inadequate to the needs of scientific inquiry. After all, the sources and causes of innovation remain mysterious, and will almost certainly always be so. That means we must constantly be thinking about and experimenting with new ways of financing and inspiring that innovation.

Innovation is difficult to dissect. In economic terms it resides inside the heads of people—people who, individually, possess different ways of seeing problems and imagining solutions. People's perspectives are accompanied by ways of searching for solutions to problems, something called heuristics. When confronted with a problem, people encode their personal perspectives and then apply their particular heuristics to envision new, possibly better, solutions.

Successful innovators obviously possess creative perspectives and productive heuristics—think Thomas Edison and his multiplicity of inventions—yet thirty copies of Edison working as a team may be no better than one. In contrast, a diverse team of individual innovators may on average use fewer heuristics each but collectively know more. When a diverse team applies those diverse heuristics, the effects can be superadditive. James Watson plus Francis Crick were together far more impressive than either working alone.

On a far larger scale, one reason for Silicon Valley's success is surely its abundance of bright engineers from different academic disciplines and from almost every corner of the globe. Collectively, they out innovate other technology hot spots that boast equal brainpower but less diversity.

Government funding of science should take this diversity calculation into account when allocating budgets. Government spending on science today is in effect a giant hedge fund. Despite the huge potential payoffs, this hedge fund won't emerge from the private sector because too often the payoffs aren't appropriable. The Naismith family made little from the invention of basketball, for example, although the world gained immeasurably.

As with any hedge fund, effective government funding of science requires that lots of money get tossed around. Some investments will yield little, while others will produce enormous dividends. This portfolio metaphor for scientific funding leads to an intuition that diversity has value, that basic scientific research should be allocated to diverse research projects. And that intuition is correct—diversity does provide portfolio insurance—yet the value of diversity goes far beyond mere portfolio effects. Diversity can produce surprising leaps.

A breakthrough in one domain can be combined with a breakthrough in another to produce even deeper knowledge. For example, research on disease transmission by epidemiologists helps us understand the spread of disease. And research by computational social scientists on how to construct large-scale simulations of societies, and the ways that citizens interact, helps us understand how markets work and how economies collapse. When we combine just these two breakthroughs, we're able to construct a third, in the form of high-fidelity computational models that can tell us when and how to intervene to minimize the medical and economic impacts of an epidemic.

In short, the mathematics of innovation shows that one plus one often equals three.

The government can encourage scientific diversity in four ways. First, it can encourage interdisciplinary research through programs such as the National Science Foundation's Integrative Graduate Education and Research Traineeship initiative, or IGERT, which funds PhD students in novel, interdisciplinary programs. The diversity of study of the IGERT program breaks through the current incentive structures of the modern academy, which reward progress within disciplines. Well-placed and sufficiently large IGERT carrots can provide incentives for scholars to step out of the comfort of their home departments to work with interdisciplinary teams.

Second, the government can continue to support scholars from underrepresented groups. Fewer than two in 100 PhDs in physics are African American, and fewer than two in ten are women. Yet we know from biology, psychology, and economics that the inclusion of women and minorities not only changes the questions being asked within a discipline, but also changes how those questions are answered.

Third, funding must loosen up, and not just the purse strings. Government grants, be they from the National Science Foundation or the National Institutes of Health, often require perfect scores from multiple referees. This tends to bias awards in favor of safer, more conservative grants. Fear of failure, which is unavoidable given that future funding prospects generally depend on past successes, exacerbates the tendency toward more mainstream research. But innovation won't be produced by tinkering on the margins of existing approaches.

In its golden era of innovation, Bell Labs actually demanded a certain failure rate. Too much success signified a lack of experimentation. Breakthroughs in science come from someone seeing a problem in a new way. But that is unlikely to happen if grant renewals require success at every step.

Fourth, the government can make commitments to big problems—finding a clean source of energy, colonizing Mars, curing cancer, eradicating poverty and disease. Big problems create diversity by shifting attention away from techniques and toward solutions. Big problems almost always unpack into lots of smaller problems, each of which requires its own way of thinking. Thus, they spawn multiple approaches. Also, big problems, such as the space program, make science focal and fun, encouraging more people, and more diverse people, to choose science as a career.

Many of the problems we face today are complex. They consist of diverse, dynamically interconnected parts. Certainly climate change, epidemics, terrorism, and poverty fit into that category.

Other problems are just plain difficult: finding a workable form of fusion, understanding protein folding, and curing diseases. We won't solve these problems, the difficult or the complex ones, with current modest levels of funding for well-established routes of inquiry. We need more funding, but equally we need more ways of thinking. The funding of science should reflect those twin needs by rewarding diverse thinkers, funding interdisciplinary research, broadening the pool of scholars, and focusing attention on big problems.

INTEGRATING ETHICS

Sometimes the way a problem is posed is a setup for going down the wrong path. So despite the common refrain surrounding certain technologies that we have a choice between science and ethics, it's best not to assume that science is necessarily "against" ethics or that the choice is truly one of either/or. Our modern understanding of science, after all, is built on the basic ideas of the Enlightenment, including values of openness and transparency, and the notion that ideas must stand or fall on their own merits.

It might even be that the science-or-ethics formulation inadvertently gives the impression that scientists don't need to worry about ethics, or that ethicists can whistle in the dark regardless of the evidence. Here are some lessons from history and emerging scientific developments that show the importance of tracking our values while in pursuit of new facts.

Neuroscience Takes Aim at Military Intelligence

N JULY 2008 A COMMITTEE OF THE UNITED STATES SENATE REVEALED THAT, beginning in 2002, Guantanamo Bay interrogators had based their methods partly on a chart that appeared in a 1957 paper prepared by an Air Force social scientist. The chart represented a summary of the types of coercive measures used by Chinese communist interrogators against American POW's during the Korean War, causing them to make a number of false confessions of U.S. war crimes. These measures fell under headings that included "sleep deprivation," "prolonged constraint," and "exposure." At the time, consternation about the effectiveness of the Chinese methods led to vague but deep-seated fears of brainwashing.

The irony that information gleaned from circumstances involving the torture of American soldiers more than fifty years ago could be used against detainees in the U.S. war on terror was not lost on opponents of the Bush administration's policies. Yet the questions raised by this incident are but the tip of the iceberg of a much larger set of questions relating to twenty-first-century neural and behavioral science, the role of that science in national intelligence operations, and the scientific community's degree of responsibility for how that role unfolds.

The American intelligence establishment's infamous Cold War forays into experiments with hallucinogens and other mind-altering processes can be attributed in part to worries that Eastern bloc communist governments were ahead of the intelligence game and less likely to respect ethical constraints than the West. One scenario was that an American nuclear physicist with a high security clearance attending a conference abroad could be invited to an apparently innocent meal and made "indiscreet" with LSD. The CIA's MKUltra and other top-secret experimental programs were among the excesses that were revealed by government investigations in the mid-1970s.

Some believe that the United States continues to pay the price for these excesses—or, perhaps, for their revelation—even to the extent of blaming intelligence failures prior to the 9/11 attacks on the resultant weakening of the CIA's covert-operations capacity. What does seem indisputable is that the American intelligence community's internal expertise on matters of the brain and behavior is not what it was in the 1950s. Some of the world's top scientists, including Harvard University psychologist Henry Murray and Harvard/Massachusetts General Hospital's Henry Beecher, were then deeply engaged advisers to military and civilian intelligence agencies, continuing relationships that began during World War II.

It is difficult to know whether that expertise actually improved performance of national-security activities. But it seems certain that without such links, the nation risks losing its edge in this important arena. And indeed, there are indications today that American intelligence capacity has been somewhat degraded by a failure to integrate the best and most up-to-date academic work in fields like anthropology and cognitive science.

For example, cultural insensitivity is often cited as one of the reasons for the early failure of the occupation of Iraq. Soldiers often had to learn on their own the nuances of communication with locals. Sometimes failures to make intentions clear, as for example in passing through checkpoints, may have had tragic consequences due to cultural variations in the meaning of seemingly simple hand gestures for "proceed" and "halt."

U.S. Secretary of Defense Robert Gates, formerly president of Texas A&M University who has served as deputy director of the CIA, recently announced an initiative called the Minerva Consortium. Minerva is intended to provide a group of universities with funding to assist the Department of Defense in areas such as Chinese military and technology studies, perspectives on terrorism in Iraq and elsewhere, religious and ideological studies, and "new disciplines" including history, anthropology, sociology, and evolutionary psychology. In an April 14, 2008, speech he also elaborated at length on the history of complicated relationships between the defense establishment and academic anthropology. With candor that surprised some, Gates conceded, "Understanding the traditions, motivations, and languages of other parts of the world has not always been a strong suit of the United States. It was a problem during the Cold War, and remains a problem." He associated these difficulties with a tension that has persisted between the American military and academia since the Vietnam War era.

As the American military establishment reaches out anew to the university system, what will be the reaction? Although Gates urged rapprochement and cited several institutions that have created special programs for injured veterans who might not otherwise qualify for admission, this is a far cry from the kinds of close relationships that characterized the World War II and postwar era. In an era in which federal funding for medical science has diminished in real terms, American academic research leaders have more motivation than patriotism alone to take an interest in a lucrative new government funding source for the generally undersupported "soft" social sciences.

Yet much of what they have to offer could prove profound. Neuroscientists cite evidence, for example, that cultural differences may extend even to the way members of different groups process information, and that these differences are measurable. If it is true that scientific understanding of culture and group dynamics has deepened in the past half century, necessitating renewed interest on the part of security officials, how much more must that be the case for the scientific study of the brain and its functions. Neuroscience conferences now rival the world's largest medical meetings, bringing together a wide range of disciplines, from psycholinguists to electron microscopists. Even taking into account the hyperbole that seems to accompany much modern science, it's a good bet that our basic understanding of the brain and its functions is on an impressive growth path.

Smart defense planners are well aware of the buzz about the brain. During the summer of 2008, a U.S. National Research Council committee, of which ths writer was a member, issued a report on *Emerging Cognitive Neuroscience and Related Technologies.* The bland title belies the fact that this was apparently the first time the American intelligence community sought systematic and rather public advice on the future of brain research from a group of scientists and academics. For longtime followers of the intersection between science and security, this turn of events was intriguing.

The charge for the NRC was in part to "review the current state of today's work in neurophysiology and cognitive/neural science, select the manners in which this work could be of interest to security professionals, and trends for future warfighting applications that may warrant continued analysis and tracking by the intelligence community." The expert committee that was recruited to consider these issues was asked to focus in particular on work that might be done in selected other countries.

Over about a year and a half, the committee's deliberations congealed around several themes reflected in the final report. In each case the national intelligence implications were paramount: Could new devices overcome challenges to the detection of psychological states and intentions, so that deliberate deception could be identified far more reliably than with traditional lie detectors? In what directions might neurologically active drugs take us, perhaps as tools for cognitive enhancement? What if computational biology leads to intelligent machines, or aids in creating human-machine systems that combine and leverage the abilities of both? What are the prospects for acquiring intelligence on cognitive neuroscience developments that might be accomplished by our competitors and adversaries? How will culture and ethics influence the hypotheses that other countries might find of interest and their willingness to engage in human experiments? In a tribute, perhaps, to the possibility of a meeting of the minds between intelligence officials and academics, none of the report is classified. Predictions of specific technical breakthroughs are kept to a minimum, and where disagreements arose among committee members, as for example over the genuine prospects for advances in lie detection, they are duly recorded.

As important as any of the report's particular findings or recommendations is the fact that it lays a predicate for a series of critical social questions that we must face even if the more extravagant expectations for emerging neuroscience are not realized. To address some of those questions, we may be able to rely on familiar territory, like risk-benefit assessments for clinical research involving implants or more powerful magnets. But as advances in imaging and computing establish more reliable correlations between neural activity and behavior, privacy limits may stretch. Governments may have to amend international conventions to establish whether interrogations of prisoners may include investigation of psychological states through real-time measurements of neural function and spatial localization. As the report notes, unethical applications of neuroscience should be at the forefront of our concern.

The environment of modern science is far more public and transparent than ever. Simultaneously, the role of applied science in the manner and methods of political violence—for nonstate as well as state actors—seems to be accelerating. Scientists are therefore under far more pressure to assess what "professional ethics" means as they participate in addressing great societal and political challenges. Perhaps only nuclear physicists have previously faced such scrutiny, though biologists, too, have in the past several years been drawn into relatively novel problems like publication concerning biological weapons.

Moreover, in the twenty-first century, the community of science is less localized than ever. Modern communications and publication technologies make data-sharing vastly more efficient. Obstacles to international collegial exchange have largely fallen, with the significant exceptions of government control of Web access or the granting of visas. In an abstract sense the culture of science has always resisted political boundaries, a cosmopolitanism that has long earned the suspicion of jealous dictators like Hitler and Stalin. But now a globalized scientific community is a functional reality. Inevitably, that community will be obliged to assess its cultural role and political responsibilities in a far more focused fashion than has previously been the case.

Uncle Sam
to Enemy:
What's on
Your Mind?

I N SPORTS STADIUMS, ON COLLEGE CAMPUSES, AND IN FERTILITY CLINICS, enhancement has become a hot topic. Athletes are shooting EPO. Students are cramming on Ritalin. And would-be parents are choosing (and before long may be designing) the in vitro fertilization embryos of their screaming-bundle dreams.

The growing availability of various enhancement technologies has amounted to a full-employment act for bioethicists, who are busily asking what is right and wrong about people's desire to be better than good. But enhancement is just one side of that conundrum. As described in a remarkable report recently released by the National Research Council (NRC), there is an equally fascinating movement afoot by the nation's military and intelligence communities to develop methods for *degrading* performance—of enemy soldiers and terrorists, of course.

The goal is to apply the latest findings in cognitive science to the job of weakening people's wills and perhaps even reading their minds, or at least deciphering their intentions before they act, all in the name of democracy and freedom.

And you thought a shot of growth hormone in a baseball player's butt was the biggest doping crisis facing the nation.

The 151-page NRC report was commissioned by the Defense Intelligence Agency. It was released with spylike discretion in the fall of 2008—no press conference or major media blast—bearing a title too bland to be bland by accident: *Emerging Cognitive Neuroscience and Related Technologies*. Taking a new angle on the truism that the brain is associated with intelligence, it argues that the U.S. intelligence community must do a better job of keeping up with advances in the neurosciences. And to some extent it echoes conventional thinking about enhancement, noting that there is a large and quickly growing market in drugs and other products that can boost physical strength and cognitive performance, which can benefit not just bicyclists and weight lifters but also U.S. forces in battle.

"In the future," the report notes, "as soldiers prepare for conflict, [the Department of Defense] may call on the neurophysiology community to assist in maintaining the warfighting superiority of the United States. Commanders will ask how they can make their troops learn faster. How can they increase the speed with which their soldiers process large amounts of information quickly and accurately? How can the neurosciences help soldiers to make the correct decision in the difficult environment of wartime operations?"

Let's ignore for now how this message contradicts what is perhaps the biggest antidoping argument tossed around by Olympic commentators, namely, that a focus on enhancement "sends the wrong message" about drugs to our nation's kids. Suffice it to say that the link between warfare and sports runs deep, and it is hard to imagine a society that honors enhancement on the battlefield but truly shuns it in the sports arena.

But even more interesting than the report's discussion of enhancement for national security is its discussion of the emerging market in brain-targeted, performance-*degrading* techniques. Some experiments, it turns out, suggest that magnetic beams could be used to induce seizures in people, a tempting addition to the military's armamentarium. Scientists are also eyeing chemicals that can blur thinking or undermine an enemy's willpower. As engineers perfect aerosolized delivery systems that can deliver these chemicals directly to the lungs (and from there, the brains) of large groups of people, the prospect of influencing the behavior of entire enemy regiments becomes real.

Indeed, in a crude way, that is exactly what Russian troops did in 2002 during the Moscow theater crisis, when they incapacitated rebels with a narcotic gas, fentanyl. But in a perfect war, the attack would be more subtle and perhaps even covert. Imagine a warrior who suddenly, inexplicably, is simply not in the mood for war.

"Although conflict has many aspects, one that warfighters and policy makers often talk about is the motivation to fight, which undoubtedly has its origins in the brain and is reflected in peripheral neurophysiological processes," the NRC report notes. "So one question would be, 'How can we disrupt the enemy's motivation to fight?' Other questions raised by controlling the mind: 'How can we make people trust us more?' 'What if we could help the brain to remove fear or pain?' 'Is there a way to make the enemy obey our commands?' . . . As cognitive neuroscience and related technologies become more pervasive, using technology for nefarious purposes becomes easier."

Suddenly, the idea of winning the enemy's hearts and minds becomes weirdly biochemical.

The report acknowledges that this approach to dealing with international squabbles is likely to stir some controversy. "The brain is viewed as the organ most associated with personal identity," it says, so "there is sure to be enormous societal interest in any prospective manipulation of neural processes."

But cognitive warfare is potentially "more humane" than old-fashioned warfare—"pills instead of bullets," in the report's words—making this a likely growth industry, the NRC concludes. And if nothing else, it suggests, the United States should be a leader in the field so that if our enemies develop such weapons then American soldiers can have the best defenses available.

"The fear that this approach to fighting war might be developed will be justification for developing countermeasures to possible cognitive weapons. This escalation might lead to innovations that could cause this market area to expand rapidly. Tests would need to be developed to determine if a soldier had been harmed by a cognitive weapon. And there would be a need for a prophylactic of some sort."

Moreover, the report says, with perhaps a subliminal nod to Abu Ghraib, "The concept of torture could also be altered by products in this market. It is possible that someday there could be a technique developed to extract information from a prisoner that does not have any lasting side effects."

This is important not only because photos of hooded prisoners with wires attached to them are embarrassing, but also because, as noted in the report, one of the real drivers of torture today is scientists' ongoing failure to develop reliable means of determining whether someone is lying or telling the truth. Of course, neuroscience can cut both ways, helping torturers extract information but also helping captives resist. In what the NRC report acknowledges may be a "far-fetched" but not necessarily crazy example, one can imagine soldiers getting Botox injections before a mission to prevent their facial expressions from giving away information if they get captured and interrogated.

The NRC is probably correct that these and similar avenues of scientific inquiry deserve better monitoring than is now underway in the secretive hallways of American intelligence agencies. No nation wants to get caught by surprise by a fancy new cognitive weapon that makes its soldiers suddenly willing to settle for the bronze medal in World War III.

But where and when will the discussions of human rights, privacy, and sovereignty come in? How do these nascent technologies fit into existing international conventions on warfare, on the treatment of prisoners, and on civil and political rights, medical experimentation, and informed consent? Surely Congress deserves to know what methods are to be used when it makes the precipitous decision to go to war. War was hell long before it was all in our heads, but it could get worse. It's time to update the oversight of these novel neural applications, to better balance the nation's legitimate security needs with the freedoms spelled out in our Constitution and in international declarations of rights.

Biopolitics and the Quest for Perfection

"Some day we will realize that the prime duty, the inescapable duty of the good citizen of the right type is to leave his blood behind him in the world, and that we have no business to perpetuate citizens of the wrong type."

WHEN THE ABOVE QUOTE WAS PRESENTED NOT LONG AGO TO AN undergraduate class, students attributed it to Adolf Hitler, and none guessed its actual author, Theodore Roosevelt. Seen through the prism of the Holocaust, "progressive eugenics" seems more like an unimaginable oxymoron rather than the mainstream science policy of social progress that it was to so many early-twentieth-century reformers. Although Margaret Sanger did not apply her views to specific groups and abhorred Nazism, "planned parenthood" included the opportunity to reduce the transmission of undesirable traits through sterilization; in some cases, mental institutions sterilized retarded and mentally ill patients. And the deep imprint of these policies lives on: several states have only recently issued formal apologies for all those thousands of lesser types they sterilized. Eugenic public-health practices rival Prohibition as the greatest success-turned-disaster in the history of American progressivism—all the more so because its history has been largely forgotten.

The old eugenics movement has become a favorite of conservative commentary, but the commentators in question seem not to know more than the bumper-sticker history. In fact, progressives and conservatives favored eugenics; the most vigorous critics of eugenics were themselves progressives; and after World War II conservatives (who detested President Franklin Roosevelt and the New Deal) were distressed at the bad odor their movement had come under in the wake of Hitler's murderous racism and longed for the day that eugenics would be restored. Perhaps the most important source of support of eugenics research for more than seventy years has been the Pioneer Fund, which the Southern Poverty Law Center has identified as a hate group.

Nonetheless, progressivism connotes an aggressive commitment to social improvement. This impulse lay behind the enthusiasm for eugenics of many science-oriented progressives, amid the rush of excitement about the social implications of Darwinism. Today, there is similar excitement about the promise of applied biology, with the important difference that there is now vastly greater understanding of underlying mechanisms, a raft of diagnostic capabilities, some capacity to manipulate genetic endowment, and the prospect of much greater control ahead. It is not just a question of who should give birth, but how. Such technologies promise to make great progress against genetic disease and birth defects. But if ensuring the predominance of the "best types" is not the goal, then what is? What is the modern progressive view of biology? Considering their history, this is not a problem for progressives to take lightly.

Set against the progressive conundrum is a flurry of thinking on the right, particularly in neoconservative circles, about the ethical implications of applied biology. The University of Chicago's Leon Kass has been writing about the issue for decades—it was he who, in the early 1970s, first began serious inquiry into the ethics of human cloning—and it was little surprise to see him and his compatriots dominate President George W. Bush's Council on Bioethics in the early twenty-first century. By and large, however, Kass's circle has concerned itself with the dark portents of the modern life sciences. But does it offer practical guidance to those who see at least some progressive potential in biology?

Although conservative, or more precisely neoconservative, thinking is held to be in acute disarray after decades of setting the policy agenda, on the issue of bioethics, neoconservative writers seem more coherent and comprehensive than progressives. In the lead are think tanks, academics, and, perhaps most visibly, several influential voices who served on the Council on Bioethics, including Kass. As opposed to both traditional conservatives, who have little to say about values in science, and religiously oriented conservatives, whose trepidations are familiar to anyone who has seen *Inherit the Wind*, the neoconservative critics of modern biology have a clear and straightforward message, one deeply informed by the experience of the cruelties the previous century wrought, as well as by ancient wisdom. In a word, that message is hubris. They fear that while previous episodes of Promethean ambition have had dire but reversible consequences, emerging biotechnologies are so powerful that, by putting the nature of humanity in fallible and perhaps malicious hands, they threaten the very foundations of human dignity. The precise consequences are not always stated—and may not be predictable—but we all know how the road to hell is paved.

On the left, matters are more muddled. At the extremes, some activists fear that the vulnerable will either be exploited or left behind by those who have access to genetic improvement, while transhumanists welcome the opportunity the new technologies present to deliberately shape the next stage of human evolution. In general, though, especially among classically oriented liberals, the tendency has been to depreciate the uniqueness of genetic interventions, whether the negative eugenics of genetic screening or the positive eugenics of selection for desirable traits, in favor of a class-based economic analysis. The familiar argument is that tennis lessons, math tutoring, and college-admission tests already seek to improve individuals or otherwise sort out the superior and lesser types, advantages closely tied to economic class. Even creepy appeals for the purchase of eggs from Ivy League undergraduates with certain scholastic and physical credentials are not, in the standard liberal account, different in kind from old-fashioned upper-class trawling for a "suitable" mate.

According to this story line, the limits of interventions, genetic or not, are only reached if they inhibit a child's autonomy-based right to an "open future," a life direction that has not been determined by others. In effect, this is the left's response to the misdeeds of its eugenic past. Some also hold the explicitly libertarian view that scientific inquiry is a form of speech, and that it is therefore entitled to the usual protections against censorship. They decry the alarmist, science-fiction predictions used by those they view as antiscience, and they urge more care in hewing closer to what is reasonable and away from worst-case scenarios. By and large, this moderate left leans toward toughening the regulatory regime while wanting to protect scientific freedom. Such classically liberal positions are widely held, but they seem, to their critics at least, more a willingness to let whim, professional ambition, and market forces determine the course of humanity than an earnest attempt to come to terms with deep moral challenges. And, needless to say, they tell us little about what to do in the case that, in fact, one of the worst-case scenarios becomes reality.

Those progressives who believe these liberal views of biotechnology are inadequate include groups who identify with the green movement, whose philosophical roots are therefore kin to European leftists like Ulrich Beck. They fear a future dominated by wealthy families who can afford "designer babies," whose expensive prenatal alterations give them an added edge over their poorer fellow humans, further driving a wedge between the haves and the have-nots. And, despite the different philosophical presuppositions of these left-wing commentators from the New Right, their ultimate concern about the biotechnological threat to humanity is quite similar. Still others, including many on the left and the right (erstwhile neocon Francis Fukuyama is an interesting case in point), see no practical alternative to a regulatory regime, in spite of their misgivings about the prospects that regulation can adequately cope with what may be barely perceptible long-term trends rather than short-term risks.

These categories, however, only capture some progressives; by and large, the movement has yet to grapple with the overall issue. The most noteworthy attempt, Bill Clinton's National Bioethics Advisory Commission, reached a consensus that human reproductive cloning should not be permitted, due to the risks to the fetus and the mother. On a philosophical level, this was something of a dodge: had it not been for the known problems reproductive cloning presents mammals, it is not clear that the Clinton commission would have agreed on anything. Compared with the sometimes apocalyptic but nevertheless stridently serious language of neoconservative bioethics, such risk-benefit analysis seems a green-eyeshade approach to deep philosophical problems.

In the absence of a more philosophically thick progressive alternative, the President's Council on Bioethics clearly set the agenda for such discussions in the first Bush term. Created in the wake of, and primarily in response to, the controversy over the use of human embryos in stem-cell research, the council struck many bioethicists—a largely progressive fraternity—as a gift to right-wing ideologues and a missed intellectual opportunity. But at the end of the day, while the council's leadership did not accomplish all that the president's conservative base may have wished, it did give secular, conservative bioethical voices far more visibility than they had before, and in doing so paved the way for a coherent conservative approach to bioethical questions.

Considering this background there should be space for an ameliorative middle. One member of the council, Michael Sandel, has made such an effort. Sandel's sensitivity to the stem-cell issue has been enriched by his close relationships to Kass and distinguished biologist Douglas Melton, a sober advocate and practitioner of human embryonic stem-cell research. Indeed, his position between those two informs much of his writing on the issue. Where Melton is currently researching replacement pancreatic cells (an important weapon in the fight against type 1 diabetes) and has developed his own embryonic stem-cell lines that are, thanks to Bush, ineligible for federal research funding, Kass steadfastly takes the opposite position, warning that such research indulges scientific arrogance and pushes the moral envelope beyond societal acceptance.

Sandel's strategy involves depreciating the standard autonomy-based liberal view as lacking the depth the subject demands. After all, Jürgen Habermas argues that direct genetic manipulation fails the liberal test precisely because it violates the principles of autonomy and equality: parents can shape their children's futures to an unacceptable degree. Sandel agrees but thinks more is needed to understand the transgression of hyperparenting. Drawing on theologian (and fellow council member) William May's notion of the "unbidden" as a special lesson of parenthood, he contends that if parents are in a position to choose more traits for their children, they will be excessively responsible for their children's fate. If children fall short, then, it is because their parents failed to make the right investment in some constituent of their design. No one would be self-made or expected to be. My limitations would be due to someone else's failure to outfit me completely or correctly. Sandel would therefore seem to draw a bright line between research with a therapeutic aim and cosmetic (if that's the correct word) procedures meant to enhance a perfectly normal child.

Sandel does, however, betray some bias toward Kassian pessimism by arguing that a society in which the contingency of talents is lost would also be one in which we will lose sympathy for those who are not so favorably endowed: "Perfect genetic control would erode the actual solidarity that arises when men and women reflect on the contingency of their talents and fortunes." This perhaps goes too far; human solidarity has long been in short supply, in spite of the pervasively accidental nature of our abilities. Would matters be markedly worse in a world rife with genetic remedies? Why wouldn't we be more, rather than less, inclined to human solidarity when it is so clear that one's inadequacies are not necessary or permanent, that our flaws are biologically based rather than the result of weakness of will, or the evil eye? If all of us could, in principle, be genetically "improved," those less fortunately placed might elicit our sympathy as having been failed by those (their parents, their genetic engineers, or whoever) responsible for their design.

Of course, this whole way of thinking smacks of science fiction more than science, and it is fair to ask if public policy should be developed in light of anxieties that follow from only one of many possible distant futures. When I was growing up, we were all supposed to be zipping around in flying cars by now. The technical capacity is there, but Jetson-mobiles just don't pay off. The same might turn out to be true of much genetic manipulation. Who can tell?

Has Sandel found a middle way? His may ultimately be the least-worst position available. While neoconservatives find the entire drift of the new biology disquieting, preferring to put matters in the hands of wise counselors rather than ambitious scientists and voracious biocapitalists, Sandel is more disposed to seek a natural balance in the context of particular cases. He plays Aristotle to Kass's Plato. Nevertheless, at the end of the day, the implications of Sandel's traditionalism are not so far from that of Kass and other neoconservatives. Neither would leave science or the industrial interests behind them to their own devices. Both are deeply suspicious of the rise of consumer genetics.

As Aristotle warns the reader in his *Ethics*, one should not expect more precision in the analysis than the subject matter permits. Nonetheless, by integrating May's ideas of the "unbidden" and "giftedness" into a novel anti-liberal framework, Sandel poses an important challenge to contemporary progressives who have failed to grasp the importance of the emerging biopolitics. He helps us appreciate the central point that neoconservatives have championed: if the new biology is indeed our destiny, we need to take it seriously, anticipate the consequences, and learn from the prior life-denying eugenic embrace.

Stem Cells and the Begetting of Moral Milestones

THE CONTROVERSY OVER HUMAN EMBRYONIC STEM (hES) CELL RESEARCH has long been a cipher for society's feelings about scientists' increasing control over biology, in particular reproduction. And nothing has added more complexity to that controversy than the November 2007 publication of two papers reporting the successful reprogramming of human skin cells into a pluripotent, or embryonic stem cell-like, state.

These so-called induced pluripotent stem (iPS) cells, when measured against hES cells, exhibited the same relevant properties, including the expression of various telltale proteins and the ability to differentiate into the three primary kinds of cells that go on to produce an embryo. Such cells could, in theory, be differentiated into any of the more than 200 cell types in the human body, including cells specific to particular patients and to particular diseases. Remarkably, scientists had to activate only four of the thousands of genes in those skin cells to accomplish the feat.

These developments were immediately hailed by opponents of embryonic stem-cell research as marking the end of the hES cell debate—a solution at last to the problem of respecting human life while simultaneously advancing science. But even the scientists who led the work were quick to counter that the new cell-reprogramming technique, if it is to advance, will for some time have to be done in conjunction with continuing studies of conventional hES cell lines. Indeed, if opponents do not accept this scientific and political reality they may end up impeding progress of the very science that could actually ameliorate the moral dilemma they perceive.

More ironic still, iPS-cell techniques seem to present ethical questions of still greater scope and complexity than the embryonic stem-cell research that has excited such vigorous opposition. Thus the development of induced pluripotency of adult skin cells may in fact prove to be a dramatic demonstration of the ultimate result some most fear: human beings' mastery over their own basic biology.

Bush administration officials, under longstanding pressure for having placed severe restrictions on federal funding of hES-cell research back in 2001, had followed closely and hopefully the work that would eventually climax in the creation of iPS cells. Preliminary efforts to make iPS cells, led by the University of Wisconsin's Jamie Thomson and Kyoto's Shinya Yamanaka, were cited in a January 2007 White House document that made the case for alternatives to hES-cell research. When the now-celebrated papers were finally published in October 2007, the White House was eager to take credit, noting that some of the Wisconsin work had been supported by a grant from the National Institutes of Health.

Exaggeration is the elixir of partisan political advantage. The fact of NIH support was undeniable, but the administration went more than one bridge too far when it asserted that its tight policy on the funding of embryonic stem-cell research actually caused stem-cell biologists to seek alternatives to hES cells. First, the Bush-imposed limitations quite clearly could not account for the Yamanaka group's accomplishment in Japan. Second, the idea that adult cells could be reprogrammed was a subject of almost immediate speculation when hES cells were first isolated in 1998, because such an approach would yield cells compatible with the immune system of the individual providing the original material. Thus it became an early goal of laboratory work even though it was thought to be a much more elusive target than proved to be the case. Third, to the extent that administration backers had advocated for alternative stem-cell research methods, primarily through the President's Council on Bioethics, they had emphasized other approaches—ones that never got much scientific traction and, in some cases, were even received with deep moral reservations among anti-abortion activists.

Finally, and perhaps most provocatively, it is likely that, if anything, the inability of the NIH to fund research on a variety of hES cell lines from 2001 to 2007 actually impaired biologists' understanding of the properties that typify pluripotency. Thomson himself has alluded to the delayed science that was a consequence of the political controversy.

As Thomson and Yamanaka have pointed out, there are at least three reasons why hES cell lines will continue to be needed. First, the iPS cells are not ready for use in research or therapy. The current technique for reprogramming skin cells involves retroviruses and cancer-causing genes. Thus other reprogramming techniques will have to be devised, perhaps involving chemicals or other molecules. Although there is reason for confidence that this can be accomplished, in the meantime if knowledge of stem-cell biology is to flourish, work on hES cell lines will be required.

Second, it is possible that iPS cells will not in fact be wholly interchangeable with hES cells. Honesty requires acknowledging that no one can know how much any sort of pluripotent cell can do or not do as a function of its inherent characteristics. The reprogramming process itself may effect limitations that render the treated cells less than fully potent and therefore unable to model naturally occurring cells of specific types. The field is just too new to be sure. Therefore, again, if the field is to grow while these answers become available—probably in a matter of years—research on the only other source of pluripotent cells should continue.

Third, efforts to turn iPS cells into disease-specific cells (say, for drug development), and tissue-specific cells (for the particular patient who was the source of the original skin cell), may require hES cells as a comparator indefinitely. To see why this may be the case, recall that the way that the induced pluripotency was confirmed in the recent papers was by comparing their properties with hES cells. Similarly, in order to know whether the reprogrammed skin cells are doing what is desired in a particular experiment or laboratory preparation, their activity will need to be compared with that of an hES cell undergoing the same differentiation process as the iPS cell. This sort of comparison may be required for a very long time, at least until biologists and regulators have sufficient confidence in their understanding of the adequacy of the reprogramming process that induced pluripotency.

Induced pluripotency is exciting for scientists because the laboratory processes involved appear to be simpler than those used for embryos and hES cells. But these advantages also point to the possibilities to which Yamanaka has darkly alluded: "We are presenting new ethical issues, maybe worse ones. . . ."

Consider, for example, that a central theme of opposition to hES-cell research is a deep anxiety about the implications of biotechnology for the future of human dignity. Although such concerns are a central motivation for modern bioethics as a whole, many bioethicists who oppose embryonic stem-cell research lack confidence in the prospect for regulation of applied biology so that it does not transgress consensus moral norms. For them, regulatory frameworks are suspect because history suggests that government tends to be in the thrall of the science establishment, which in turn creates "accommodationist," pseudoethical schemes that only appear to set moral limits to science. When necessary, the science establishment obtains the services of academic bioethicists who are already inclined to give their blessing to scientists' wishes to pursue their goals in a largely unconstrained manner, according to these critics.

But scientists and bioethicists alike will have to confront the new possibilities for artificial human reproduction inherent in the rapidly advancing stem-cell biology. Already a British group has reported that it has coaxed human female embryonic stem cells to develop into cells with some of the essential qualities of sperm. Suppose one were to pursue an attempt to transform a diploid cell (a body cell with all forty-six chromosomes) into a haploid gamete (a sperm or egg cell with only twenty-three chromosomes). This might involve first expelling half the genetic complement (as has apparently already been done in mouse cells), and then treating the remainder with factors that are required for gametic processes.

It seems that opponents of embryonic stem-cell research who celebrated the advent of iPS cells have not grasped that the cellular reprogramming technique actually aggravates their greatest concerns about the power of modern biology. For if skin cells can be reprogrammed to become pluripotent and then differentiated into specific somatic cell types, they may also be differentiated into germ (sex) cells.

Since male iPS cells have both X and Y chromosomes, they could be reprogrammed to sperm and eggs. These iPS-derived sperm and eggs could then be used in standard in vitro fertilization procedures. Notably, couples with an infertile male partner may be able to obtain sperm that could then be transferred to the woman's uterus. The resulting infant would have virtually the full complement of DNA of both members of the couple, though whether the male can be called the father in the traditional biological sense will be a matter of debate. Alternatively, it may be possible for a gay male couple to obtain an oocyte derived from the skin cell of one member of the couple, which could then be combined with the sperm from the second man and the embryo brought to term through the services of a gestational or surrogate mother. The resulting child would be genetically related to both men. Lacking a Y chromosome, a lesbian couple would not be able to reproduce in this way. However, if the genes sort differently during the formation of each of the gametes, there would be grave risks for any resulting embryo. It should not be necessary to elaborate on the extraordinary ethical and social questions that would be raised by such developments. In an overview of the issues raised by pluripotent stem cells, the Hinxton Group, an international consortium on stem-cell ethics and law, urged caution in any new regulatory regime that might be stimulated by these questions: "In the case of PSC-derived gametes, as with all science, it is important to target policy specifically to those dimensions of the research or its applications that have proved to be unacceptable, and that these policies be proportionate to the magnitude of what is morally at stake."

But that is not the end of the story. If iPS-derived germ cells are in the offing, then so are blastomeres, the cells that constitute an embryo at its very earliest stages. To turn a diploid cell into a blastomere one might either use the induced germ cell in a process of parthenogenesis or spermatogenesis, or introduce factors that skip the gamete stage and turn the iPS cell directly into a blastomere. Thus will come to pass the most astonishing and disorienting result of all: modern stem-cell biology will at that point have made every cell of our body a potential embryo.

All of these scenarios tread the dangerous territory between science and science fiction. The genetic resorting that would take place through several steps of reprogramming from adult cells to iPS cells to gametic cells would almost surely make it too dangerous to attempt human reproduction, so as in the case of reproductive cloning, issues of risk would have to be dealt with before more profound ethical issues would need to be addressed. Yet how many stem-cell biologists, including Thomson and Yamanaka, predicted that reprogramming would be accomplished so quickly?

A more plausible scenario for the use of iPS cells to produce a genetic twin of the cell donor has already been demonstrated in mice by scientists at Advanced Cell Technology, a Worcester, Massachusetts, biotech company. "We now have a working technology whereby anyone, young or old, fertile or infertile, straight or gay can pass on their genes to a child by using just a few skin cells," a company official said. Moreover, the official added, "the bizarre thing is that the Catholic Church and other traditional stem-cell opponents think this technology is great when in reality it could in the end become one of their biggest nightmares. . . . It is quite possible that the real legacy of this whole new programming technology is that it will be introducing the era of designer babies."

Widespread appreciation of this technical reality could have profound effects on the divisive abortion debate, but in what directions? There seem to be at least several distinct possibilities, all of which may co-exist. The first and most likely short-term possibility is that prolife groups will split between those that wish to ban such procedures as antithetical to the natural process of conception and those that find it an acceptable alternative along the lines of in vitro fertilization. A second, more extreme, result and far less likely result would be that the human embryo in its early, disorganized state prior to, say, the appearance of the primitive streak (roughly around fourteen days) comes to be seen as no more than another clump of cells. Eventually, something like the traditional view still reflected in Islam, Judaism, and most Protestant denominations may once again be accepted even by those who once held a more elevated view of the early embryo. A third possible outcome of the advent of iPS-derived embryos, and one that is perhaps the most distant, is that a growing proportion of the public comes to view the tissues and organs that compose the human body as the remarkable systems they are, rich with life and the potential for independent life. Cults that worship every cell and even every sloughed cell can be imagined.

What is clear is that our society is unprepared for breakthroughs in the life sciences that we can foresee just over the horizon. For some, the new dawn of mastery over our own biology that will follow from the technology of induced pluripotency will seem like a cruel joke and confirm their worst fears. Some may even be reminded of the myth of Prometheus, whose punishment for stealing fire from the gods and sharing it with humans was to be tethered to a rock where his liver was consumed by an eagle. Thus we may conclude that, though humans may suffer for their knowledge, neither will it consume them, for the liver happens to be the only organ in the human body capable of complete regeneration, a definitive property of pluripotent cells.

UPDATING THE WAY WE EDUCATE

Considering how long human beings have been teaching kids, you might think we'd do a better job. Part of the problem is that as soon as we've figured out how to teach something, something new comes along. And some of the challenges to effective teaching are external to the educational process itself, like poverty, or poor health care, or a dearth of employment opportunities for those in a position to acquire the needed knowledge. And then there are the shockingly pointless, distracting, and phony debates about matters like evolution.

We need to get to the serious business of improving education, and therefore lifetime opportunities, in science and math at the lower-school level. Fortunately, America's higher-education infrastructure is the envy of the world, a tremendous driver of energy and creativity. But it can't pull the wagon alone, and remediation is not the answer.

STEMing the Tide

ACCORDING TO A CONTINUING STREAM OF REPORTS AND WHITE PAPERS from eminent think tanks and government agencies across the country, the United States faces a shortage of technical talent that threatens our future competitiveness. This shortage, it is said, arises largely from inadequate kindergarten-through-twelfth-grade public-school education in science, technology, engineering, and math, the so-called STEM curriculum. But this perception of dearth and mediocrity, though widespread and widely accepted in political and policy circles, ignores the most important issue undermining U.S. science. Reforms are urgently needed, but not the ones offered up by those who are focused on education.

Prominent labor economists who have examined the problem from a different perspective argue that poor STEM education isn't the problem at all. In fact, they believe there are far too many qualified student-scientists. Rather, it's the perverse financial incentives that American society (and specifically the U.S. government) offer wannabe American scientists that lie at the heart of our nation's science and technology competitiveness crisis.

At first glance, those arguing that we face a shortage of scientists appear to make some valid points. It's true, for example, that fewer top-ranking native-born, white male students—the demographic that long provided the bulk of the nation's technical and research professionals—are pursuing graduate studies in science. Ditto that a growing percentage of the scientists in training at the nation's universities are foreign-born. And the average performance of U.S. K-12 students on international standards is indeed undistinguished.

But these facts do not add up to the crisis that critics describe. Rather, according to a number of distinguished economists, they reflect a labor market gone seriously awry. In the first place, average test scores tell nothing about the supply of students capable of becoming scientists. The youngsters who have the option of taking that path are not the ones who are average for their age group, but are outstanding, and the U.S. produces these standouts in large numbers. One frequently cited international comparison, for example, shows that the United States had far more top-performing science students than any other nation tested, as well as a big lead in the number of top-performing readers, as documented by Hal Salzman of the Urban Institute and B. Lindsay Lowell of Georgetown University. Americans also are surpassed only by Japan in the number of top scorers in math.

What pulls down the U.S. average is not an overall deficit but the very poor performance of students at the bottom, largely products of inferior schools serving poor, minority communities. These disparities are a national disgrace that should be corrected, which in turn would result in an even more qualified and more diverse pool of talent to improve our nation's competitiveness. But our poor test scores say nothing about the quality of America's best schools, which rank among the world's finest, and the pipeline of qualified students available for careers in science.

The top performers from those excellent schools then proceed to study at some of the world's best universities, also conveniently located in the United States. Professors at these universities encourage the most promising to continue on for science PhDs, in preparation for careers as academic researchers. The students who take this advice hope for satisfying careers resembling those their senior professors have enjoyed, pursuing their best ideas as independent researchers, heading labs amply supported by federal funding, and enjoying job stability and comfortable upper-middle-class incomes as faculty members in secure, tenured positions.

But the world that nurtured today's senior professors, with PhDs earned in four years and appointments as faculty members and lab heads in their 20s, has vanished. What the great majority of today's young scientists find instead is a penurious decade or more working in university labs, first as graduate students and then as postdoctoral researchers earning a trainee wage comparable to what a new liberal-arts BA graduate makes.

Their search for the faculty post essential to starting their own academic research careers overwhelmingly ends in frustration, as they futilely compete for every advertised faculty opening against hundreds of other qualified applicants—all of whom sport good degrees and lists of publications from their graduate and postdoc years. The odds that a young PhD will ever land a faculty job at any four-year institution are now less than 25 percent, and at the kind of research university where top-notch science is done, well under 15 percent.

Across the United States, therefore, professors are watching in horror as many of their brightest undergraduates eschew science graduate study in favor of medical, law, or business school. These students don't reject science because they're bad at math, but because they're good at it. Anyone bright enough to get a science PhD is bright enough to run the numbers showing that an average of seven years of graduate school, followed by five or more postdoc years, followed by long odds against getting the job one was ostensibly preparing for, add up to a lousy investment.

For foreigners, however, especially those from developing countries, grad school or a postdoc in America is exceedingly enticing. Why? Because the virtually unlimited visas that universities can supply make such training an otherwise largely unobtainable ticket into the country.

Labor economists, including Paula Stephan of Georgia State University and Richard Freeman of Harvard University, believe this excess of young American scientists unable to start their academic careers results from "the perverse funding structure of science graduate education," as fellow labor economist Michael Teitelbaum of the Alfred P. Sloan Foundation put it in congressional testimony.

Research grants to individual professors from the National Institutes of Health, National Science Foundation, and other agencies finance the great bulk of graduate students and postdocs. To get the grants and renewals needed to keep their labs going, professors must produce steady streams of journal articles. That, in turn, encourages them to have as many grad students and postdocs as they can possibly afford to do the bench work. This highly skilled cheap labor makes American research very economical, but produces as a by-product intense pressure on the system to absorb the waves of incoming students into jobs that simply do not exist.

Proponents of the hypothesis that U.S. science is suffering from a lack of scientists, which in turn stems from inadequate early education, counter that low unemployment among early career scientists proves there is no glut. In fact, however, the postdoc pool, now numbering possibly 90,000, of which more than half are foreign-born, constitutes a disguised unemployment pool—a holding tank for overqualified "trainees" who are, to their frustration, frozen out of the market. In fact, the United States annually produces about 30,000 new science and engineering PhDs, about 18,000 of them American-born, while faculty openings at research universities in the most glutted fields have been estimated to hover in the mere hundreds.

The tiny minority who do land research-based faculty jobs have spent so much time training that, in biomedical science, for example, they average forty-two years of age when they finally launch their independent-research careers by winning their first competitive federal grant. At that age, scientists of previous generations—Albert Einstein, Marshall Nirenberg, Thomas Cech—were collecting Nobel Prizes for discoveries made in their twenties.

"I try to keep my best undergraduates away from my postdocs," one professor confided, because meeting them would reveal what really lies ahead on the grad-school track. But talented young Americans would flock to science study if it offered them the kind of career opportunities that previous generations enjoyed. Instead of a needless general overhaul of K-12 education, or an increase in graduate fellowships, which would only make things worse, the United States needs to overhaul what Brown University biochemistry chair Susan Gerbi calls the "pyramid paradigm"—a reference to the classic pyramid schemes that promise so much but leave participants high and dry.

Instead of paying universities to use grad students and postdocs as very smart migrant laborers, the U.S. government needs a funding structure that provides large numbers of them a solid career ladder into the life that so many were implicitly promised. The jobs on that ladder need not offer them the glamorous financial rewards that come with corporate law, medical specialization, or investment banking; science offers intellectual riches so much more dazzling than money, it has long enticed the ablest young Americans to accept more modest remuneration in exchange for the chance to do great research. But the futures we provide to the young people we ask to devote their lives and talents to learning and doing science must match those other careers in providing at least a reasonable likelihood that with hard work and devotion they can attain their goal.

At present, the United States does not give them that opportunity. One way to start doing so would be to structure funding to encourage universities and lab chiefs to create jobs for permanent staff scientists who receive professional-level salaries, benefits, and status within the university. Another could be requiring universities to limit the graduate-student and postdoc positions they create to the number of people who could reasonably be expected to find career-level employment after they leave their professors' labs. Another could be requiring universities and lab chiefs to track their grad-school and postdoc alumni and report on their employment experience to new applicants, as professional and business schools routinely do.

When the nation once again provides its young scientists a decent shot at the life they hope for, our best youth will race to answer science's call.

Of Fun, Games, and Finals

THE NO CHILD LEFT BEHIND ACT (NCLB) HAS CERTAINLY SPAWNED SOME creative thinking over the years, though much of it not the kind that the act's crafters had in mind. Many states have jiggered their tests to obscure the failure of poorly performing schools. Others have attempted to opt out of the program. And every state has complained about the lack of available funding to support the act's goals. The process has been wasteful, confusing to students, and has failed to produce the information and tools that education enterprises so badly need.

It's time for a new approach, to be launched in two stages. First we should engage in a national debate about the expertise students will need to prosper in the twenty-first century. Then we should settle on how best to measure their progress, drawing as much as possible on the same kind of creativity that has long been finding itself at home with students: the interactive methods used in computer games. These methods represent some of the most powerful ways to test newly acquired skills. But to understand why, we must first recognize why our current testing procedures are so thoroughly outdated.

Here's the first counterintuitive fact about how NCLB has been working: despite all the complaints about the numerous tests it mandates, the problem is not too many tests but too few. High-stakes, standardized tests are an artifact of a mass-production model of education imposed on students and schools out of necessity during the last century. Traditional tests measure performance in situations that will seldom, if ever, occur in an actual job. Someone trained to solve problems working in isolation, with no access to reference material and no ability to consult experts, is largely useless in today's economy.

But consider the ideal classroom scenario: an instructor able to spend plenty of time with individual students, constantly challenging them, asking probing questions, and presenting increasingly complex problems and exercises tailored to the needs of each student. By the time a test is taken, the student should have run through the material enough times that success is virtually assured.

These powerful methods aren't used in standard classrooms for two obvious reasons: they're unaffordable, and we continue to think of the classroom as our forebears did 200 years ago. Yet a solution is available from an unexpected source—computer games.

The average U.S. teenage boy spends about fourteen hours a week glued to computer games. Most adults can't imagine how the lessons of Super Mario could be applied to high school science or history, but consider that a good game captures and holds a player's attention with a series of compelling goals, each slightly beyond the player's current abilities. A great game draws players into what designers call the flow, where they will try, fail, and try again, working for hours to master the skills needed to win.

What's striking, of course, is that they're also being continuously tested. Tests are an integral part of winning, and players accept that they will fail before they master the skills needed to move on. If you keep crashing your simulated aircraft, you know that you've got to work harder. Winning at the most advanced levels of game play requires players to draw on a huge body of knowledge and experience.

Winning many games, moreover, often requires more than mastery of specific skills. It requires precisely the skills that the Partnership for 21st Century Skills recently reported are in greatest demand in today's economy: gathering evidence, evaluating options, making decisions under conditions of uncertainty, and—in the case of multiplayer games—working effectively as a member of a team.

The U.S. Department of Defense, which unlike most organizations is unembarrassed about having its employees play (okay, war) games, has come to appreciate the power of simulation-based games to teach and test individuals and teams. Department officials have convincing evidence that skills acquired through simulations translate into performance in the field.

Simulation-based instruction can reproduce the complexity, confusion, and tension of field conditions so faithfully that the success a soldier gains in a good simulation translates directly into reliable performance during his or her first real combat experience. This powerful transfer from simulation to practice has also been demonstrated for pilots and surgeons. Surely, it's possible to create challenges in biology, history, or engineering that can capture and hold a middle-school student's attention.

Building software to teach and test complex skills is expensive. Several billion dollars were invested and lost in education technologies toward the end of the dot-com boom a decade ago, and investors have been wary of this sector ever since. Schools and universities are a notoriously poor market for innovations, in part because of an understandable reluctance to take risks with unproven approaches. But as a result, an enormous opportunity is being lost.

The federal government should fill this gap by funding basic-science research, development, testing, and evaluation of game-based educational techniques that can then be picked up and commercialized by private investors. Imagine the new twist for parents across the country: "Charlie, how many times do I have to tell you? It's late! Stop doing your homework!"

Knuckle-Walking to the Final Exam

ONE OF THE COROLLARIES OF THE THEORY OF EVOLUTION IS THAT THE fat lady never sings. The whole point about evolution is that it is a process, ever ongoing and in flux. A species does not stop evolving unless and until it goes extinct. So perhaps it should not be a surprise that in this day and age, more than a century after the basic tenets of evolutionary theory were spelled out and despite mountains of more recent scientific evidence to back it up, the vaunted theory still finds itself beleaguered by opponents.

Most recently the venue was Louisiana, where the legislature in 2008 passed a law that sought to undercut the teaching of evolution in the state's schools. State Senate Bill 733, ironically named the Louisiana Science Education Act, called upon the state's Board of Elementary and Secondary Education to "create and foster" a school environment that promotes "objective discussion of scientific theories being studied including, but not limited to, evolution, the origins of life, global warming, and human cloning."

The move follows others in school districts in states such as Texas (where schools were directed to teach the "strengths and weaknesses" of the theory of evolution), Kansas, and Pennsylvania, each of which, in its own way, has sought to sow the seeds of doubt in students' minds with regard to the evidence for evolution. Many of those efforts, including the Louisiana gambit, have overemphasized the word "theory," a word that to many unschooled in the scientific definition of that term sounds rather more provisional than it really is. In science, "theory" is the word used for explanations of phenomena that are supported by overwhelming evidence. Consider that the scientific explanation for why objects do not float helter-skelter into the upper atmosphere is the theory of gravity. No one doubts that gravity exists just because it is "merely" a theory. Will Louisiana's science teachers before long come under a mandate to teach the arguments for and against the theory of gravity?

Setting aside for now some of the technical glitches in the Louisiana legislation (global warming and human cloning are not theories at all—a concept that Louisiana's legislators *should* have learned in school—so it is difficult to understand how teachers are to fulfill the new requirement to discuss those "theories"), the new law is but the latest incarnation of an effort by indefatigable religious fundamentalists to spike science classes with a dose of creationism, which relies on supernatural explanations for the origin of Earth and the life that lives on it. In particular it reflects the vibrancy of the intelligent-design movement, which proposes a pseudoscientific approach to understanding biological diversity and complexity that has been faulted repeatedly by scientists and neutral judges alike.

If you thought this battle had already been won, you'd be wrong, though your error would be understandable. 'Twas five days before Christmas 2005, when the United States District Court for the Middle District of Pennsylvania delivered its highly anticipated decision in *Kitzmiller v. Dover Area School District*. At issue was the legality of a 2004 Dover Area School District decision to inform all students that they should "keep an open mind" about evolution and to encourage students to peruse *Of Pandas and People*, which the school district gamely referred to as "a reference book," to gain an understanding of a competing view of how life came to be, known as intelligent design.

Judge John E. Jones III did not pull his punches. He found that the testimony of school board members who favored the teaching of intelligent design in the schools "was marked by selective memories and outright lies under oath." He labeled intelligent design as "a religious alternative masquerading as a scientific theory." And he stated plainly that intelligent design was "the progeny of creationism." That's important, because in a previous ruling the U.S. Supreme Court had banned the teaching of creationism in science classes.

Indeed, Jones highlighted a secret game plan written by leaders of the intelligent-design movement that made clear the real goal of these various academic battles. The "Five Year Strategic Plan Summary," known to fundamentalist insiders as the "Wedge Document," states that the movement's goal is to replace science as currently taught and practiced with "theistic and Christian science." The group's "governing goals," according to this document, are to "defeat scientific materialism" and to "replace materialistic explanations with the theistic understanding that nature and human beings are created by God."

At one point in the trial, Judge Jones dissected what he called the "historical pedigree" of the book *Of Pandas and People*. Not only is it published by a group registered with the Internal Revenue Service as a religious, Christian organization, he noted, but a look at the various versions it went through over years of editing reveals something rather amazing. Early versions of the manuscript, written before the Supreme Court's creationism decision, refer throughout to creationism. Later edits, completed in 1987 after the court's ruling, are virtually identical but for the substitution of the words "intelligent design" wherever the word "creationism" had appeared. "This compelling evidence strongly supports Plaintiffs' assertion that ID is creationism re-labeled," Jones concluded.

Now America's schools find themselves under assault by yet another relabeling of creationism, which this time is making its cowardly run for the academic goal line under the linguistic guises of "strengths and weaknesses" and "objective discussion of scientific theories."

Do we really need more "objective discussions" of gravity or the atomic theory? Are any of the proponents of these antiscience movements willing to stand up for their purported doubts by, say, stepping off a cliff or warming themselves by the glow of some nuclear waste? C'mon, they're only theories.

One would be hard-pressed today to find scientists beating down church doors to teach rationalism to parishioners in their pews. Yet supernaturalists seem unwilling to refrain from foisting their beliefs on kids in science classes. School policies and curricula are primarily local in this country, and on the whole that is a good thing. Yet history has shown that when it comes to larger issues of fairness and pedagogical rigor, it is appropriate for the federal government to step in and set standards—whether the prompt is integration, gender fairness in sports, or basic academic standards as spelled out in the No Child Left Behind Act.

Perhaps it is time for the federal government itself to speak up in favor of the facts that explain how we humans got so smart. There is no reason why God, faith, and even black magic and supernaturalism should not be topics of hot discussion in philosophy, religion, and culture classes. But federal tax dollars should not be supporting schools that persist in teaching myths in science classes.

EMPOWERING
THE PUBLIC

The scientists' old lament that "If only people understood more about science . . ." does not have the punch it once seemed to have. For one thing, science is becoming ever more complex, if anything growing more out of reach to the layman than it was even a few years ago. For another, given the increasingly large array of commitments and short-ages of time most people have today, it simply doesn't make sense to expect that the average person can become much of an expert in sci-entific affairs.

Yet that does not mean that the public is fated to be powerless when it comes to participating in societal discussions or decisions that embrace some element of science or technology. New tools (of sci-ence!) are enabling everyday people to learn enough of what they need to know when they need to know it. Novel media outlets, from YouTube to fake news shows to video games, are finding the "sweet spots" in the American psyche that make people interested in and even attracted to certain things scientific, without the burdensome overlay of frank teaching or testing. And a growing number of scien-tists themselves are willing to enter the public- and government-policy fray, to stir their prescient, pocket-protectorish perspectives into the rest of the American cultural gemisch.

As these final writings demonstrate, the urge to innovate and to reach upward and outward—a trait seemingly native to human exis-tence itself—is blossoming in countless different ways as technology breaks down the boundaries of biology and culture. We are partici-pants in a remarkable and only marginally controlled experiment in communication, collaboration, and, ultimately, self-knowledge, enabled by our unique capacity to figure out how the world works. There is no stopping the process, but that does not mean we are powerless to choose our path. This is the responsibility, and the indescribable thrill, of science next.

People Are
from Earth,
Scientists
Are from...

FUNDING FOR UNIVERSITY-BASED SCIENCE AND ENGINEERING RESEARCH HAS lagged behind the rate of inflation for more than two years now. That's a first, and one that raises a complicated question: Where, exactly, does science stand in America today? Is it respected? Disdained? Or just ignored?

On the positive side, Americans express strong confidence in the leaders of the scientific community. Among important institutions of society, polls suggest, only leaders of the military are better trusted. (In case you're wondering, the other end of the trust scale is populated with journalists and members of Congress, who are generally considered—to use the scientific term—slime.) Americans also claim to be very interested in new scientific discoveries and developments. According to the National Science Foundation (NSF), surveys conducted each year from 2001 to 2006 found that from 83 percent to 87 percent of Americans have either "a lot" or "some" interest in new scientific discoveries.

Obviously, though, this isn't the full picture. The same Americans who express such confidence in the leaders of science have a lot of trouble naming any of them. When polled in late 2007 and asked to name scientific role models, the best that Americans could come up with were the names of people who were either not scientists or not alive: Bill Gates, Al Gore, Benjamin Franklin, and Albert Einstein. Moreover, while many people claim to be very interested in new scientific discoveries, they're even more interested in other things. According to 2006 data from the NSF, only 15 percent of the public follows science news "very closely." Science ranked behind ten other news subjects in terms of people's interest. And there is some evidence that the ranking of science vis-à-vis other news subjects is slipping of late. Certainly the number of dedicated science pages and science sections in newspapers has been steadily declining.

Nowhere are the ambivalent feelings that Americans seem to have for science and scientists more apparent than in the public-policy arena, where scientific information gets stirred into the stew of political debate. On the one hand, everybody—left, center, and even Christian right—claims that science lies on their side. The perception is basically universal that it is good and advantageous to appear to be proscience and to be able to claim that your side is informed by science.

As soon as you get into the details of what politicians or advocates are actually claiming, however, things quickly get murky. The science often fails to support their assertions, and the proscience aura quickly dissipates under scrutiny—supplanted by opportunism or, in some cases, outright cynicism and manipulation. Consider the arguments about how much mercury we should let polluters spew into our atmosphere, or the carbon emissions and food-price increases produced by making ethanol from corn. Suddenly all that trustworthy science finds itself outweighed and outgunned by lobbyists who are more beholden to finance than fact.

And it's not just the lobbyists who are so fickle. The same goes for ordinary members of the public, whose strong respect for and interest in science often proves to be rather skin-deep. If you put it to the test— by asking the public to, say, take sides in a perceived conflict between science and their religious beliefs—suddenly science doesn't fare very well. According to the NSF, when Americans reject the evidence for evolution or the big bang (which they do far more frequently than citizens in many other countries), it's not, for the most part, because they don't understand the basics of what the science says. Rather, it's that they don't let science win out in competition with other things that are important to them—in this case, religion.

This is not to suggest that people should abandon their most deeply held beliefs whenever a scientist comes along with a different take on things. But when it comes to deciding important issues of public policy in a pluralistic nation whose constitution draws a bright line between church and state, it's not enough to be a fair-weather friend to science. Daniel Yankelovich, the social scientist and pollster, perhaps put it best:

> Science has reached greater heights of sophistication and productivity, while the gap between science and public life has grown ever larger and more dangerous, to an extent that now poses a serious threat to our future. We need to understand the causes of the divide between science and society and to explore ways of narrowing the gap so that the voice of science can exert a more direct and constructive influence on the policy decisions that shape our future.

And of course, it is not only deep-seated religious beliefs that help feed this divide. More often than not, less lofty and more expendable aspects of human nature, such as superstition and rumormongering, are the culprits undermining rationality. Consider the recent test-drive of the world's most powerful subatomic particle accelerator, the Large Hadron Collider, or LHC, built by Europe's CERN (the European Organization for Nuclear Research), and the ripples of fear—which ultimately spread around the world—that its activation might in some way destroy the world.

The alleged means by which it might do so were numerous, the most common being that the machine's high-speed proton collisions could create tiny black holes that would grow to engulf the Earth and everything around it.

Scientists repeatedly dismissed the concerns. For one thing, they noted, cosmic ray collisions with the Earth's atmosphere are happening all the time and are much akin to the kinds of collisions the LHC is designed to produce under controlled conditions, yet we're all still here. Such explanations notwithstanding, and despite all the respect that the pollsters tell us these scientist-explainers have, the apocalyptic fears—and even a few lawsuits aimed at halting the project—persisted. People, it seems, just can't help themselves. (And of course it doesn't help that Hollywood has trained us to think of big, strange science projects as having an inherent tendency to go badly awry.)

The 2006 public outcry over Pluto's status as a planet offers another emotional example of the disconnect between what people say about their respect for science and how they actually feel about the things that scientists conclude. Recall that 2006 was the year that the International Astronomical Union voted to demote Pluto to the newly created rank of "dwarf planet." That decision by a relatively small group of scientists, made on the strict basis of technical considerations, prompted a global backlash that continues to this day. In such cases, the public, which hasn't been monitoring developments in science, is suddenly shocked to hear what is going on. And the scientists, who haven't been monitoring the public, are just as surprised at the backlash.

And yet these are critical moments for the world's scientific community, centrally because it's so hard to get science on the public radar to begin with. When it finally does occur, and you've got the public's attention, you don't want it squandered over something petty, like the Pluto issue, or something silly, like fears that the Large Hadron Collider will make us all cease to exist. The good news is that such developments spark dramatic levels of interest in science; the bad news is that they're highly negative encounters.

So how should we deal with such situations? It's a difficult problem for which both the public and scientists ought to take some responsibility. But scientists in particular, it seems, ought to be capable of anticipating public reactions, monitoring worrisome sentiments and misinformation, and seeking to engage long before things reach the boiling point.

In fact, that's precisely what scientists and the federal government are trying to do with nanotechnology, the science of engineered microscopic particles. There's a lurking worry, fanned by Michael Crichton's novel *Prey* and other chatter, that this form of research, like the LHC, could unleash some sort of apocalypse. But the National Nanotechnology Initiative, the federal coordinating body for nanotech research, has engaged scholars, including social scientists and even philosophers, to study and anticipate such concerns as part of an effort to encourage public participation in and, yes, public acceptance of, nanotechnology. Of course, nanotechnology may well pose some risks, and they should not be ignored. But all told, it seems a good thing that in this case the fear of misunderstanding and backlash has spurred smart attempts to reach out in advance to stakeholders and inform them about what nanotechnology is and is not, and what it can and cannot do.

There's outreach from CERN, too. Several safety studies have been undertaken to respond to public fears and, perhaps even more effective, there's the hilarious "Large Hadron Rap" video, which has educated and entertained millions of people on YouTube. Still, these approaches have been more reactive than proactive, and you don't have to be a physicist to know that it is a lot harder to catch up with a train that has already left the station than it is to catch one that is not yet barreling down the tracks. So as we wait for those subatomic particles to finally collide, let's not forget to study collisions between scientists and the public as well. In those human encounters, as in the LHC, there are strong and poorly understood forces at work. Forces with lots of potential energy, and that really do matter.

Transcending
the Public-Trust
Monologue

I N 2004 THE GENETICS AND PUBLIC POLICY CENTER, A NONPROFIT BASED in Washington, D.C., fielded a survey of more than 4,000 U.S. residents to plumb their attitudes about new genetic technologies. The results were less than gratifying: more than 40 percent of those who responded said they did not trust scientists "to put society's interest above their personal goals."

The roots of this uneasy relationship lie in the reliance that the science and technology community has long placed in various "deficit models" of interaction with the public. The basic assumption behind these models is that there is a linear progression from public education to public understanding to public support, and that this progression will inevitably cultivate a public wildly enthusiastic about research. But this model of scientific engagement with the public obviously isn't working.

Lately, all manner of ways to "involve" the public in science policy and practice have cropped up, mostly around oversight of emerging technologies such as synthetic biology, nanotechnology, and human genetics. Scientific associations are developing centers devoted to public engagement in science, funding agencies have created sweeping mandates for collecting public input on research, and research-performing institutions are hosting community meetings and science cafes about their work. But one might wonder: Are these new organizations truly "engaging" the public?

In a nutshell, an erosion of public trust that began as a trickle of doubt about radiation safety and pesticides has grown in recent decades to program-threatening uprisings against a wide range of new technologies, from genetically altered "Frankenfoods" to the mythical "grey goo" that some believe will emerge from nanotechnology.

Initially, the "deficit" in question was framed as an information deficit: if only laypeople knew what scientists did, goes this line of thought, they would support the agendas of the scientific establishment. Since World War II, the science community has been operating under this information-deficit model, built on a one-way flow of information from experts to the public with almost no information flowing back the other way. This model drove communication of science and technology for the last half of the twentieth century, despite a glaringly inconvenient truth: neither public support for research nor scientific literacy increased significantly in that time.

More recently, the information-deficit model has begun to be reframed as an attitudinal deficit. To know us is to love us, runs the mantra of this public-understanding school of science-society interaction. Having realized the practical futility—if not the questionable ethics—of making every layperson a lay scientist, the public-understanding model contents itself with pursuing public appreciation, emphasizing the benefits of science to society without worrying unduly about how much science the public actually understands. The end goal hasn't changed— increased public support of science and technology—even if the methods used to get there and the metrics used to define success are different. The direction of information flow remains the same as well: top-down from the scientist or engineer to the public.

The asymmetric communications practices embodied by the scientific-literacy and public-understanding movements cultivate scientists who resist ceding any level of control of the science-policy agenda to nonscientists, a view neatly encapsulated by a quote from a series of scientist interviews conducted at the Genetics and Public Policy Center a few years ago:

> I don't think that the general uninformed public should have a say, because I think there's a danger. There tends to be a huge amount of information you need in order to understand. It sounds really paternalistic, but I think this process should not be influenced too much by just the plain general uninformed public.

This wariness is reciprocal in the twenty-first century, as U.K.-based communications researcher Martin Bauer and his colleagues noted in the journal *Public Understanding of Science* last year: "Mistrust on the part of scientific actors is returned in kind by the public." Negative public attitudes like those, they say, are in turn viewed by scientists as proof that "a deficient public is not to be trusted" to provide uncritical support for the scientific enterprise. And the wheel of miscommunication goes round and round.

Clearly, something needs to change in the science-public landscape. Writing in the journal *Science* in 2003, the chief executive of the American Association for the Advancement of Science, Alan Leshner, summarized the problem eloquently: "Simply trying to educate the public about specific science-based issues is not working. . . . We need to move beyond what too often has been seen as a paternalistic stance. We need to engage the public in a more open and honest bidirectional dialogue about science and technology."

Indeed, research-performing institutions increasingly say they have traded in their old, top-down models of science literacy and public understanding for the new buzzwords of "public consultation" and "public engagement." But the philosophy behind consultation and engagement seems, on closer inspection, not to have changed much at all. Many scientists expect consultation and engagement to cultivate a public more supportive of science as planned by, performed by, and promoted by scientists—despite the fact that neither consultation nor engagement have been rigorously evaluated to see if these goals are reasonable or even possible. And even if they turn out to be measurably effective in meeting some articulated goal, are they affordable enough to deploy? Neither consultation nor real engagement can be done on the cheap.

What, then, can consultation or engagement do for us? This "participatory turn" in science-society relations, as Harvard scholar Sheila Jasanoff terms it, ostensibly focuses on regular dialogue (two-way, symmetrical communication), transparency of the decision- and policy-making process, and meaningful incorporation of public input into these processes. On paper, the goal of these two-way, participatory models is mutual satisfaction of both parties, the research enterprise and the public—if not with every specific outcome then at least with the overall relationship that exists between them.

Key dimensions of this dialogue are negotiation, compromise, and mutual accommodation. It places a premium on long-term relationship building with all of the strategic publics: research participants, the media, regulators, community leaders, policymakers, and others. These emerging models offer promise for scientists and the public to engage each other more fully and productively, although the promise is as yet only tantalizing, and not yet tempered or informed by much scrutiny from social-science research.

In fact, the dearth of evaluative research on these experiments in engagement stems in large part from the fact that, despite all the talk, very little real experimentation is going on. In practice, much communication currently passed off as public consultation and engagement is still one-way, expert-to-layperson information delivery, albeit in different settings like cafés scientifiques, public meetings, and town halls. Research organizations have been quite adept at putting together well-rehearsed, tightly scripted opportunities for "public input"—but with no institutionalized mechanisms for truly reflecting the public's input in policy-deliberation processes or in the actual construction of new policies. In fact, one gets the not-so-subtle impression that these engagement events are being held with the hope of staving off public dissatisfaction, or providing just enough semblance of listening to public concerns that the natives don't get so restless that they revolt.

The endgame of public engagement should be real empowerment: creating a meaningful mechanism for public input to be heard far enough upstream in science and technology policymaking and program development to influence decisions. It is not about the scientific elite making a decision and then staging public events to move the public toward agreeing with that desired outcome. It is about empowering lay citizens to learn all they want about pending science or policy issues (not what scientists believe they need to know to weigh in), and then giving them access to deliberative processes that allow that knowledge to be questioned, applied, and melded with additional knowledge or questions gleaned from outside the scientific process.

Publican engagement is about agreeing upfront to accommodate public input politically, not just to listen and nod politely. Unlike the unidirectional and hierarchical communication that characterizes scientific-literacy and public-understanding models of science-society relations, public engagement—when practiced as an honest, iterative dialogue—does result in demonstrable shifts in knowledge and attitudes among participants. The Genetics and Public Policy Center has documented and measured these shifts during town-hall and online deliberations. Of course, the shift is not always in the direction scientists might expect or prefer. After all, public engagement is not about getting the policy you want; it's about getting the public input you need to craft sustainable policy that enjoys public confidence.

In the end, as complicated and bureaucratic as it may seem, public engagement is personal. It changes people, and not only those people who make up that grand, vague notion of "the public." Yes, the public gains knowledge, shares expertise, and reflects on how much risk that it, as a society, is willing to accept in return for a shot at the promising benefits of emerging technologies. But equally significant, public engagement can change scientists, too. This is the real mark of successful public engagement: Not simply insisting upon the public's deeper appreciation and understanding of society's preferences and values. At such times, real communication is possible, and the path to public trust can begin in earnest.

You Bought It, You Should Own It

TUCKED AWAY IN THE 2009 FEDERAL BUDGET IS A PROVISION THAT benefits the scientific community without spending a dime: a mandate that recipients of funds from the National Institutes of Health make their published research results available to the public.

The policy is the first "open-access" mandate adopted by the U.S. government, and puts teeth into the voluntary policy in place at the agency since 2005. The NIH, which supported the provision, moved quickly to implement the law, which reflects similar policies instituted by various U.S. foundations and universities and several foreign-funding agencies.

Under the new policy, publicly supported scientists—who received about $29 billion in taxpayer funding in fiscal year 2008, a figure greater than the gross domestic products of 100 countries— must deposit a copy of their research articles into the National Library of Medicine's PubMed Central database within twelve months of those articles being published in a scientific journal. PubMed Central will then make those articles available for free to the worldwide research community, as well as to the general public. Previously, the results of grantee research were available only from the journals in which that research was published—either by subscription, which can cost thousands of dollars annually, or by paying per-page fees that can quickly add up for someone trying, say, to understand a family member's diagnosis.

The new policy is notable not only for its progressive stance and the whopping amount of research it will make available to the public— NIH-funded scientists produce an estimated 80,000 published articles annually—but also for the political clash it sparked. In 2007, open-access advocates ramped up their efforts, led by the Alliance for Taxpayer Access, a letterhead coalition driven by the Scholarly Publishing and Academic Resources Coalition, a consortium of academic libraries.* In addition to rallying grassroots enthusiasm from patient advocates and others, proponents circulated a letter of support signed by twenty-six Nobel laureates, including former National Institutes of Health Director Harold Varmus.

Opponents, led by members of the Association of American Publishers, launched a countercoalition dubbed PRISM, the Partnership for Research Integrity in Science and Medicine. That group predicted that, by removing journals' monopoly control over their content, open access would take a financial toll on publishers and undermine the peer-review process that is so important to scientific journal publication. Some opponents even likened public control over federally funded results to government censorship of private publications. PRISM in turn drew ridicule from science bloggers, who criticized the group's statements as Orwellian and derided the "grassroots" group as Astroturf, since PRISM declined to list its own membership. The group also took some hits for hiring Eric Dezenhall, known in Washington as "the pit bull of PR," to develop a public relations strategy to combat open access. Even some members of the Association of American Publishers distanced themselves from the effort.

In the end, the open-access provision was signed into law—the biggest legislative victory to date for the American open-access movement and, given the vast impact of NIH funding, for advocates worldwide. The adoption of the policy will introduce more life scientists and others to self-archiving—the posting of one's own research results online for free access, already common in physics and other areas of science—than any single event to date.

Journals will have to adapt to the new business model, but are not expected to suffer many, if any, subscription cancellations. High rates of self-archiving in physics have not been shown to have caused any drop in subscription levels in that field. Publishers may, however, feel more pressure to provide value and limit price inflation, pressures the

---

*The author is a consultant and former intern for the coalition.

lucrative industry has not had to think much about in recent years. From 1986 to 2002, journal costs rose 227 percent, more than triple that period's overall rate of inflation as measured by the Consumer Price Index.

In addition to the general public, the scientific enterprise itself is sure to benefit immeasurably as researchers gain more complete access to the scientific record. Even the wealthiest research institution cannot afford to subscribe to every journal in publication. Free online access also lays the foundation to remove unnecessary permission barriers, using approaches such as the Creative Commons licenses, and to facilitate machine-assisted research via Semantic Web technologies, which can help scientists compare and contrast results from large numbers of studies.

With implementation now well underway at NIH, other research-funding agencies may find more courage to pursue open-access policies of their own. This may pave the way for a government-wide mandate akin to the Federal Research Public Access Act, floated by Senators John Cornyn (R-TX) and Joe Lieberman (ID-CT), which would require open access for all large, research-based agencies. Some may even push for stronger mandates than the NIH policy, such as the European Research Council policy released last year that halves the NIH's maximum allowable delay from twelve months to six months after publication in a journal.

Open access is a positive development for several goals of science policy: to accelerate research, control costs in higher education, and share information more effectively. The NIH public-access policy will move forward on all three fronts and pave the way for progress to come.

Making Science Sexy

PETER CALAMAI DESCRIBES HIMSELF—AND ONLY HALF JOKINGLY—AS A "grizzled veteran" of the newspaper industry. Over the course of his forty-year career, he has covered a wide range of subjects, but for the past decade Calamai was the dedicated science reporter for Canada's most widely read newspaper, the Toronto *Star*. That was until June of 2008, anyway, when, along with one-tenth of the paper's staff, he took a buyout—an all-too-common occurrence these days, as newspapers cut back on resources in the face of declining subscriptions and ad revenue thanks to competition from the Internet. Today the *Star* retains medical and environmental reporters, who of necessity do science-related writing in the course of their work. But the paper has not hired another science-centered journalist since Calamai's departure.

In fact, the treatment of science at the *Star* was shrinking long before 2008. Until a few years ago, the paper ran pages labeled "Science" in its Saturday and Sunday editions. But when those fell by the wayside, the rest of the paper didn't make up for it; Calamai could still file science feature stories, but the total column space for science coverage declined noticeably.

In a time of media-industry upheaval, the situation at the *Star* is hardly unique. A study by longtime science reporter Cristine Russell, on behalf of the Shorenstein Center on the Press, Politics, and Public Policy at Harvard University's Kennedy School, showed that large numbers of newspapers are trimming their coverage of science. From 1989 to 2005, Russell found, the number of U.S. newspapers featuring a weekly science section declined precipitously, from ninety-five to thirty-four.

It may be understandable that newspapers are cutting back on total coverage in light of the economic challenges they face. Circulation levels for major daily newspapers have been dropping 2 to 4 percent annually for several years now, and fell even more steeply in 2008— nearly 5 percent over just a six-month period, according to the Audit Bureau of Circulations. And while online ad sales have been growing, they still account for only about 10 percent of total newspaper ad revenues, and are not even close to making up for the losses of traditional print ads.

In this context, you can see why publishers would see science as an area in which cuts might be made with little fear of substantial pushback from readers. Many adults, after all, are unabashedly uninterested in science, and some even seem to take pride in having almost failed high school chemistry or physics. Still, Calamai expected some kind of protest when the *Star* killed the science section. Alas, almost no one called to complain. Sure, there were a handful of complaints— perhaps a dozen at most, he said. That compares to the hundreds and even thousands of angry letters and calls many newspapers have reported receiving in response to occasional efforts to change the lineup on their comics pages. If science's dedicated followers aren't willing to pick up the phone and raise hell when Canada's biggest newspaper cuts back on science coverage, is it any surprise that many media outlets might conclude that they, too, need not necessarily give science its due, even as they continue to devote precious ink and paper to such crowd-pleasers as the daily horoscope column?

Downgrading science coverage is a mistake, especially in these days when science and technology are emerging as such central players in our civilization's potential to extract itself from the many global problems we face. If anything, there is a professional and even moral duty on the part of the media to cover science and cover it well—and that means the *mass* media, not just the science magazines and not even just the newspapers, but the full range of broadcast media and perhaps even some of the newer entertainment venues. To ensure that this duty is fulfilled, though, will require both a renewed commitment from scientists and others to make science engaging, as well some noisemaking by consumers who appreciate the importance of the topic.

One tactic is to be preemptive. Readers in communities where a newspaper still carries a regular science section ought to write the editors and let them know they value the content. That kind of activism might also help stem a tide many science reporters complain about today: growing pressure to add part-time blogging and other duties to their already full-time job of covering science, leaving them with far less time to produce the kind of in-depth stories that can really inform, educate, and have significant policy repercussions.

But beyond efforts like these to save the standard model of science journalism, perhaps it is also time to acknowledge that, for all the principled talk about the media's obligation to cover science and other topics in the name of elevating national discourse and preserving democracy, it may simply be that some for-profit news outlets are destined to die in the new online economy. If that is the case, then perhaps some major news organizations need to consider radically different futures in order to continue the important work they do, such as becoming nonprofit 501(c)3 organizations, committed to serving as neutral, educational, public-service institutions essential to the maintenance of democracy.

In fact, there is evidence that such a trend has already begun. Witness, for example, the launch of Pro Publica, a nonprofit, investigative-journalism outfit that does not count on making money from ads or subscriptions, but regards itself as a public service and offsets some of its expenses by selling its exposés to major media outlets. Other science-news sites, some based at universities, have also begun to operate with foundation money.

But lest the whole situation start to seem dreary and hopeless, let us not forget that the word "media" encompasses a far larger universe than newspapers and magazines and educational Web sites. There is, for instance, the big, flashy, and exciting world of entertainment. And while many a scientist and science reporter may recoil at the prospect of their work getting "dumbed down" into cartoon format, there are some increasingly creative outlets available that are proving the possibility of informing and entertaining all at once.

Consider the "World Science Festival," held in New York City in 2008—a four-day three-ring, scientific circus of movie stars, particle physicists, magicians, philosophers, neuroscientists, authors, inventors, evolutionary biologists, paleontologists, composers, planetary scientists, astronomers, and a host of others representing the full spectrum of science and the humanities—including a dance troupe performing an artistic interpretation of black holes and string theory. The festival's ambitious goal: "To cultivate and sustain a general public informed by the content of science, inspired by its wonder, convinced of its value, and prepared to engage with its implications for the future."

Cofounded by Columbia University physicist Brian Greene and his wife and former ABC producer Tracy Day, the festival was a success by any number of measures. Not only were events filled to capacity, but the festival generated a host of stories in newspapers. Moreover, in very smart fashion, festival organizers did not settle for conventional science coverage. Greene, Day, and actor Alan Alda together rang the bell that closed out the NASDAQ stock exchange one day during the festival, drawing attention in economic quarters. New York Mayor Michael Bloomberg gave the festival's opening address, providing fodder for political reporters. And the festival's panels were liberally populated with stars such as *The Bourne Identity* director Doug Liman, making the event a magnet for entertainment buffs and paparazzi.

As a result, coverage of the festival was not limited to newspapers' news pages. Stories found homes in the society, style, and entertainment sections, reaching readers who might not typically find themselves reading about physics, biotechnology, or chemical reactions. Even more exciting, the festival got prime billing on ABC's *Good Morning America* (which covered the event twice and even seized the opportunity to declare science "sexy" and geek "chic") and achieved what many media mavens see as the pinnacle of pop visibility: Greene was interviewed by Stephen Colbert on Comedy Central's *The Colbert Report*, the highly popular "fake news" entertainment program. There, before an audience of more than a million viewers and in a mere five minutes, Greene was able to work around Colbert's repartee about "large-breasted alien robots" and science's besting of religion ("You guys been kicking ass since the Enlightenment," Colbert quipped) with an astonishing smorgasbord of informational tidbits about quantum physics, string theory, gravity, general relativity, and the essential nature of scientific experimentation.

"Science," Greene said, "is the greatest of adventure stories." And so it is. What's more, like Scheherazade, she always has one more story to tell, each better than the last. So here's to the next. To science next!

Acknowledgments

Science Next builds on the Center for American Progress's electronic journal of science policy, *Science Progress*. Center for American Progress CEO and president John Podesta saw the need for a venue that would enable scientists and policy experts to develop progressive solutions for the technological challenges America faces in the 21st century. We are grateful to him and the rest of the American Progress administration, board and investors.

Many of our CAP colleagues have been crucial in this venture. We want especially to recognize the brilliant work of the *Science Progress* editorial staff: managing editor Ed Paisley, assistant editor Andrew Pratt, editorial advisor Kit Batten, contributing editor Chris Mooney, and fellows assistant Mike Rugnetta. Thanks are also due to CAP's legal eagles, Debbie Fine and Maytak Chin.

Bellevue Literary Press editor Erika Goldman's enthusiasm about *Science Progress* propelled the vision for *Science Next*. She is as incisive and smart as she is fun to work with. We thank Erika and the Press for their commitment to infuse the public square with ideas that are both powerful and humane.

Most of all we are grateful to the readers of *Science Progress* on the Web and to you, the readers of *Science Next*, and to our contributors. As you will discover in these pages, we have been fortunate to attract some of the most imaginative and critical writers about science and where it ought to go next.

—JDM
RW

Contributors

Robert D. Atkinson (*Riding the Sci-Tech Express*) is Founder and President of the Information Technology and Innovation Foundation.

Gavin Baker (*You Bought It, You Should Own It*) Assistant Editor of Open Access News and an Outreach Fellow at the Scholarly Publishing and Academic Resources Coalition.

Joseph W. Bartlett, (*Don't Cry for Me, EVITA*) an advisory board member of *Science Progress*, serves as Of Counsel at Sullivan & Worcester LLP, Courtesy Professor, Johnson School of Business, Cornell University; and Founder and Chairman of VC Experts, Inc.

Sarah Bates (*Going with the Flow*) is the former Deputy Director for Policy and Outreach at Western Progress, a nonpartisan regional policy institute dedicated to advancing progressive policy solutions for the Rocky Mountain West. She has written extensively on western water and natural resources law and policy.

Beryl Lieff Benderly (*STEMing the Tide*) is a Washington, D.C. science journalist who contributes the monthly "Taken for Granted" column on labor force and early career issues to the website of *Science* magazine.

Rick Borchelt (*Transcending the Public-Trust Monologue*) is Director of Communications at the Genetics and Public Policy Center at Johns Hopkins University, which is supported by The Pew Charitable Trusts.

Vinton Cerf (*American Can Do It: Reinvening the Soviet Satellite Stimlus Plan*) is Google's vice president and chief Internet evangelist. As one of the "Fathers of the Internet," Cerf is the co-designer of the Internet's TCP/IP protocols and architecture. Cerf holds a B.S. degree in Mathematics from Stanford University and MS. and Ph.D. degrees in Computer Science from UCLA.

Elizabeth Edwards (*Foreword*) is a Senior Fellow at the Center for American Progress, where she works on health-care issues and writes occasionally for the Wonk Room, a rapid-response policy blog. She is the author of the *New York Times* bestseller *Saving Graces: Finding Solace and Strength from Friends and*

Strangers, a memoir, and is active in the Wade Edwards Foundation and a variety of other charitable efforts. A passionate advocate for children and an accomplished attorney, she has been a tireless worker on behalf of important social causes.

Maryann Feldman (*Good Enough for Government Work: A Model of Quality Management*) is the S.K. Heninger Distinguished Chair in Public Policy at the University of North Carolina, Chapel Hill.

Dan Guttman, (*Outsourcing Governance*) is the co-author of *The Shadow Government* and shared in a journalism award for the report, "Outsourcing the Pentagon." He is on the faculty at the Center for the Study of American Government at Johns Hopkins University, teaches at Peking University Law School, and is a fellow at the Tsinghua University China-America Center.

Kathy Hudson (*Transcending the Public-Trust Monologue*) is Director of the Genetics and Public Policy Center at Johns Hopkins University, which is supported by The Pew Charitable Trusts. She is also Associate Professor in the Berman Institute of Bioethics, Institute of Genetic Medicine, and the Department of Pediatrics at Johns Hopkins.

Jeremy Jacquot (*Owning Up to Our Oceans*) is a graduate student in marine environmental biology at the University of Southern California and is a freelance writer.

Tom Kalil (*Rx: Innovation Infusion*) is a former non-resident Fellow at the Center for American Progress and a former advisory board member of *Science Progress*.

Henry Kelly (*Of Fun, Games, and Finals*) is the President of the Federation of American Scientists in Washington, D.C.

Richard O. Lempert (*Securing Our Scientific Future*) is the Eric Stein distinguished University Professor of Law and Sociology *emeritus* at the University of Michigan.

Mark Lloyd (*Bringing Broadband Up to Speed*) is Vice President of Strategic Initiatives at the Leadership Conference on Civil Rights and an Affiliated Professor of Public Policy at Georgetown University.

Chris Mooney (*People Are from Earth, Scientists Are from . . .*; *Making Science Sexy*) is a contributing editor to *Science Progress* and the author of three books: *The Republican War on Science*; *Storm World: Hurricanes, Politics, and the Battle Over Global Warming*; and *Unscientific America: How Scientific Illiteracy Threatens our Future* (with Sheril Kirshenbaum).

Jonathan D. Moreno (*Time for Science to Reclaim Its Progressive Roots*; *Neuroscience Takes Aim at Military Intelligence*; *Biopolitics and the Quest for Perfection*; *Stem Cells and the Begetting of Moral Milestones*) is Editor-in-Chief of *Science Progress*, Senior Fellow at the Center for American Progress, and the David and Lyn Silfen University Professor and Professor of Medical Ethics and the History and Sociology of Science at the University of Pennsylvania.

Alan Muney (*Mom, Apple Pie, and Interoperable Electronic Medical Records*) is an Executive Director in the Blackstone Corporate Private Equity Group in New York, where he is in involved in portfolio management and monitoring of healthcare benefits. A former pediatrician, he was previously the Chief Medical Officer of Oxford Health Plans.

Scott Page (*Diversity as a Driver of Innovation*) is Leonid Hurwicz Collegiate Professor of Complex Systems, Political Science, and Economics at the University of Michigan, and external faculty at the Santa Fe Institute. He is the author of *The Difference: How the Power of Diversity Creates Better Groups, Teams, Schools, and Societies*.

Hannah Pingree (D-North Haven) (*Old States Can Learn New Traits*) is Speaker of the Maine House of Representatives.

Reece Rushing (*Stress Test: Turning to Technology for Those Bridges to "No Wear"*) is Director of Regulatory and Information Policy at the Center for American Progress.

Nancy Scola (*Fixing Our Fractured Food-Safety System*; *It Takes a Village to Issue a Patent*) is associate editor at *techPresident* and Personal Democracy Forum, as well as a freelance journalist focused on the intersections of technology, politics, and culture.

Jonathan B. Tucker (*Microbes and the Military: Preparing for the Battle of Biotech*) is a Senior Fellow specializing in biological and chemical weapons issues in the Washington, DC office of the James Martin Center for Nonproliferation Studies of the Monterey Institute of International Studies.

Jim Turner (*Good Enough for Government Work: A Model of Quality Management*) was Chief Counsel of the Committee on Science and Technology and retired in 2009 after over 30 years on Capitol Hill. He currently is working on energy policy for the National Association of State Universities and Land Grant Colleges.

Rick Weiss (*Time for Science to Reclaim Its Progressive Roots*; *Solar Rays to the Rescue*; *Prescription for a Bodily Extended Warranty*; *Your DNA's Spitting*

Image; *Uncle Sam to Enemy: What's on Your Mind?*; *Knuckle-Walking to the Final Exam*) a former *Washington Post* science journalist, is a Senior Fellow at the Center for American Progress.

Steve Woodruff (*Pruning Old Promethean Policies*) is the former Northern Rockies Deputy Director for Western Progress, a nonpartisan regional policy institute dedicated to advancing progressive policy solutions for the Rocky Mountain West.

Author Key

Index